Residential Crowding and Design

Residential Crowding and Design

Edited by
John R. Aiello
Rutgers—The State University
New Brunswick, New Jersey

and
Andrew Baum
Uniformed Services University of the Health Sciences
Bethesda, Maryland

PLENUM PRESS · NEW YORK AND LONDON

Library of Congress Cataloging in Publication Data
Main entry under title:

Residential crowding and design.

Includes index.
1. Crowding stress—Addresses, essays, lectures. 2. Personal space—Address-
es, essays, lectures. 3. Architecture—Human factors—Addresses, essays, lec-
tures. 4. Architecture—Psychological aspects—Addresses, essays, lectures.
I. Aiello, John R. II. Baum, Andrew.
HM291.R455   301.1   79-357
ISBN-13: 978-1-4613-2969-5   e-ISBN-13: 978-1-4613-2967-1
DOI: 10.1007/978-1-4613-2967-1

©1979 Plenum Press, New York
Softcover reprint of the hardcover 1st edition 1979
A Division of Plenum Publishing Corporation
227 West 17th Street, New York, N.Y. 10011

To Our Parents

Contributors

John R. Aiello • Department of Psychology, Rutgers University, New Brunswick, New Jersey

Andrew Baum • Department of Medical Psychology, Uniformed Services University, School of Medicine, Bethesda, Maryland

Lisa E. Calesnick • Department of Psychology, Trinity College, Hartford, Connecticut

Sheldon Cohen • Department of Psychology, University of Oregon, Eugene, Oregon

Verne C. Cox • Department of Psychology, University of Texas at Arlington, Arlington, Texas

Glenn E. Davis • Department of Psychology, Washington College, Chestertown, Maryland

Gary W. Evans • Program in Social Ecology, University of California, Irvine, California

Jonathan L. Freedman • Department of Psychology, Columbia University, New York, New York

Omer R. Galle • Population Research Center, University of Texas at Austin, Austin, Texas

Walter R. Gove • Department of Sociology, Vanderbilt University, Nashville, Tennessee

Charles J. Holahan • Department of Psychology, University of Texas at Austin, Austin, Texas

Paul J. Hopstock • Department of Psychology, Manhattan College, Riverdale, New York

Garvin McCain • Department of Psychology, University of Texas at Arlington, Arlington, Texas

Dennis P. McCarthy • Environmental Psychology Program, Graduate Center, City University of New York, New York, New York

Walter Ohlig • Program in Social Ecology, University of California, Irvine, California

Paul B. Paulus • Department of Psychology, University of Texas at Arlington, Arlington, Texas

Susan M. Resnick • Program in Social Ecology, University of California, Irvine, California

Judith Rodin • Department of Psychology, Yale University, New Haven, Connecticut

Susan Saegert • Environmental Psychology Program, Graduate Center, City University of New York, New York, New York

Allen Schiffenbauer • Research Division, Needham, Harper, & Steers, Inc., Chicago, Illinois

Janette K. Schkade • Department of Psychology, University of Texas at Arlington, Arlington, Texas

Donald E. Schmidt • Societal Analysis Department, General Motors Research Laboratories, Warren, Michigan

Drury R. Sherrod • Department of Psychology, Pitzer College, Claremont, California

Daniel Stokols • Program in Social Ecology, University of California, Irvine, California

Stuart Valins • Department of Psychology, State University of New York, Stony Brook, New York

Brian L. Wilcox • Department of Psychology, University of Texas at Austin, Austin, Texas

Preface

The intent of this book is threefold: (1) to summarize recent research concerned with residential crowding, (2) to present some new perspectives on this important subject, and (3) to consider design implications and recommendations that can be derived from the existing body of research. We have sought to bring together the work of many of the researchers most involved in these areas, and have asked them to go beyond their data—to present new insights into response to residential crowding and to speculate about the meaning of their work for the present and future design of residential environments. We feel that this endeavor has been successful, and that the present volume will help to advance our understanding of these issues.

The study of residential density is not new. Studies in this area were conducted by sociologists as early as the 1920s, yielding moderate correlational relationships between census tract density and various social and physical pathologies. This work, however, has been heavily criticized because it did not adequately consider confounding social structural factors, such as social class and ethnicity. The research that will be presented in the present volume represents a new generation of crowding investigation. All of the work has been conducted during the 1970s, and a range of methodological strategies have been employed in these studies. Several teams of investigators have used multiple methodologies in their examination of the processes involved in the physical concept of density and the psychological experience of crowding. More importantly, the focus of these studies has been on mediating factors as well as on the consequences of crowding.

One of these mediating influences is architectural design. The buildings that we shape in turn shape us, and some of the literature on residential crowding has addressed the effects of building design in crowded conditions. Several papers in this volume are concerned with the relationship between architectural design and high-density residential settings, reporting evidence of mediation of crowding, suggesting ways of input into design processes aimed at reducing residential crowding, and discussing the relationships among design, crowding, and social behavior. It is hoped that the issues considered in this volume help designers and practitioners as well as researchers to ask more informed questions about residential crowding and design.

We (along with others) have noted the importance of studying human response to high-density conditions in situations where they naturally occur because of the setting-specific nature of crowding experiences. Crowded residential environments were among the first environments to be studied and continue to be of great importance for researchers as well as for practitioners.

Although our primary purpose for this book is to provide resource material for professionals interested in residential crowding and design, we have attempted to keep the writing style as jargon-free as possible so that the book could also be used in the classroom. We hope that it will prove to be a useful adjunct to other texts for advanced undergraduate and graduate courses in the environment and behavior or architecture and design areas.

Initially, several of the chapters in this book were recruited for a special issue on residential crowding for Plenum's *Human Ecology* by the first editor, who was then associate editor of that journal. We soon found that it was necessary, however, to expand the breadth of our treatment of this topic so that the design implications of residential crowding research could be sufficiently developed. We would like to thank Andrew P. Vayda, then editor of *Human Ecology*, for his comments on several chapters, and Daniel Stokols, who as chairperson of a session on crowding and design at the 1976 meeting of the Environmental Design Research Association, suggested several of the contributors who have participated in this book. We would also like to express our appreciation to Seymour Weingarten, former executive editor at Plenum, and to Donna E. Thompson and Carlene S. Baum, who encouraged and facilitated this undertaking. Also, we thank Amy Green, Nan McSwain, and Margaret Wideman for their help in the preparation of the indexes. Lastly, we are grateful for the grant support that we have received in support of our work, HD 0854601, HD 0754501, 5S07 RR07087–12.

JOHN R. AIELLO
ANDREW BAUM

Contents

I

The Study of Residential Crowding

John R. Aiello and Andrew Baum

Introduction

The importance of studying crowding in primary environments, where people spend much of their time relating to others on a personal level and engaging in personally important activities, should be obvious. If we assume that high density is inevitable for at least part of the world's population and that high residential density is a problem that will not go away quickly, the reasons for studying the impact of residential crowding and the ways in which its effects can be ameliorated become clear.

Altman (1973) has recommended a place by process analysis for the study of crowding. A more recent elaboration of this setting-specific type of analysis (Karlin, Epstein, & Aiello, 1978) has noted several events that occur in crowded settings and that are responsible for evoking the label "crowded." In *residential* crowded settings close physical proximity can be considered less important than the lack of control over interpersonal interactions and congestion or resource scarcity factors, which are likely to be more salient in these settings. Accordingly, most of the studies in this volume have examined aspects of high density residential conditions from a conceptual base generated by these factors. Many have focused on density conditions that lead people to feelings of loss of control in their environment (e.g., Baum, Aiello, & Calesnick, Chapter 9; Rodin, Chapter 5; Sherrod & Cohen, Chapter 13).

Studies reported in this book differ in the levels of observation used to consider the effects of crowding, but information ranging across a number of levels is represented in the chapters that follow. For example, Galle and Gove not only present information at a demographic level (where no attention is paid to characteristics of individuals), but also supply data at a normative level by analyzing for the independent and interactive effects of such variables as residents' ethnicity, occupation, and educational level. Most of the other studies were conducted at the phenomenological level, whereby individuals label their reactions to the environments. Most of these studies use questionnaires or interviews to obtain residents' responses. Other research considers nonverbal interaction behaviors to assess social withdrawal responses. Lastly, physiological indicators of social stress, using the palmar sweat technique and cortisol levels from residents' urine samples, have been used.

Recently, Altman (1978) has observed that present-day researchers in the crowding area tend to be rather eclectic and are willing to use a variety of methodological approaches. Consistent with this reflection, several of the research reports have employed more than a single methodological strategy in their work on the effects of crowding. All of the research to be reported takes advantage of naturally occurring density levels in (primary) residential environments; some bring residents to controlled laboratory conditions, whereas the others observe or survey residents in their natural milieu.

Along with Paul Hopstock, we present an overview of previous research on residential crowding in Chapter 1. Patterns of response to crowding experienced in residential environments are reviewed in the context of cognitive/affective responses to residential crowding, stress-related physiological influences, and behavioral consequences of this stress. Five advantages of the application of field experimental methodologies to the study of residential crowding are described: (1) field experiments involving residential settings allow for the study of the effects of prolonged exposure to density; (2) longitudinal research is possible, since there is a stable subject population; (3) the opportunity to study cross-situational effects of density is provided; (4) the use of inferential statistics allows causal inferences to be drawn; and (5) ecological validity is assured by examining reactions to density in a natural environment. The wide variety of settings in which field experimental studies of crowding have been conducted are also described, along with the ways density has been varied within those settings. The chapter concludes with the presentation of a longitudinal study of human residential density conditions at Rutgers University, an example of the application of field experimental techniques to the investigation of residential crowding.

As noted in the preface early research on residential crowding began in the 1920s with sociologists attempting to determine the relationship between high levels of population density and incidence of pathology. The findings of most of these investigations suggested that density was moderately related to social pathology. However, no conclusive statements regarding the effects of density on human behavior could be made since other factors could also have accounted for the observed relationship and because the measures of density were, for the most part, undifferentiated, consisting of large geographical units. In 1972 an article by Galle, Gove, and McPherson appeared in *Science* that represented a new generation of sociological studies; it not only attempted to parcel out the effects of socioeconomic status, ethnic background, and other variables that might be related to social pathology, but also investigated various levels or types of density. The results of this study indicated that the number of people per room was highly correlated with indicators of social pathology such as mortality and juvenile delinquency. In Chapter 2, Galle and Gove present a brief restatement and clarification of the findings reported in that earlier article and address many of the methodological criticisms of the early analysis. The consistency of the findings reported in the *Science* article, based on a study that was conducted in Chicago, are also examined by extending the 1960 analysis to three other time periods: 1940, 1950, and 1970. Galle and Gove conclude this chapter with a confirmation of their previous findings. For each of the several different time periods examined, density was combined with social structure to form the major component of explained variance. Density itself accounted for a small but significant amount of variance, independent of social structural factors.

A field study of urban crowding illustrating the value of studying perceived crowding as it occurs in actual residential settings is reported by Schmidt in Chapter 3. By means of a large-scale, in-community questionnaire technique, measures of a wide variety of social-environmental conditions, personal characteristics, and attitudes toward overpopulation and development in a two-community area of Southern California were gathered along with information concerning individuals' perceptions of crowding in their residence, neighborhood, and city. Measurements of physical density of the residence, the census tract in which the residence was located, and the distance of a residence from a number of land amenities were also collected. Physical and psychological predictors of the perception of crowding are discussed at these residential, neighborhood, and city levels of analysis. Of particular interest is the finding that residential density (along with the accompanying psychological variables of privacy and the importance of spatial factors) was the strongest predictor of the perception of crowding at the residential level of analysis. Results are discussed in the context of a number of theoretical approaches to the study of human crowding, and the implications of these findings for urban design are considered.

In Chapter 4, McCarthy and Saegert examine the effects of high density in a large, low-income, public housing project. Specifically, these focus on residents' experiences of social overload and the consequences of these experiences for their responsiveness to the social environment. Structured interviews were conducted in the project, assessing residents' orientations to the housing project and to life in general, their daily experiences inside the buildings, and their social relationships with neighbors on their floors and with friends and neighbors outside of the building. Comparison of the responses of high- and low-rise tenants indicated that the former reported (a) more experiences of social overload and crowding; (b) feeling less control, privacy, and safety in the interior building spaces (e.g., hallway, lobby, elevator) outside of their apartment; (c) greater difficulty in their social relations within the building and being less socially involved with friends and relatives living outside of their building; and (d) feeling less satisfied and involved with their building and the project as a whole. These findings are interpreted as indicating that people may not adapt to stress over time. Instead, the authors suggest that perceptions of crowding may become increasingly stronger with time and the decrease in social interaction caused by social overload may extend to social relationships outside of the immediate residential environment.

A different type of approach to the study of the effects of high and chronic residential density was used in the investigation reported by Rodin in Chapter 5. Children from households of various density levels were brought to the laboratory in order to test the hypothesis that living under high density conditions results in a real or perceived inability to exercise control over one's environment. In the first of two experiments, an operant conditioning procedure (in which children responded to obtain candy) was used. Results indicated that children from high-density households were less likely than children from lower density households to exercise control over the administration of outcomes when given the opportunity to do so. In the second study, children were first asked to solve either a solvable or an unsolvable problem and then were given a second solvable problem to learn. Compared with children who lived in less dense homes, those who lived under high residential density conditions were less able to master the second solvable task when the first problem had been unsolvable. Taken together, the results of both of these studies indicate that feelings of choice and control may be limited by chronic exposure to high density. The importance of perceived uncontrollability as a critical mediating process in response to crowding is discussed.

The studies thus far have examined the effects of crowding in household settings. The remaining four chapters in this section focus on the effects of crowding in other high-density environments. Prisons, jails, and offshore drilling platforms represent field settings that are particularly well suited for the study of the effects of prolonged and inescapable high-density conditions. In Chapter 6, Cox, Paulus, McCain, and Schkade present the results of a series of investigations that have examined the effects of long-term crowding in a number of federal and state prison sites and local jails. Their overall pattern of results suggests that (a) social crowding in prison settings can produce negative emotional responses with a decreasing tolerance for crowding over time; (b) illness complaints (e.g., backaches, nausea, asthma) are greater under more dense prison conditions; and (c) high levels of social density may increase physiological indications of social stress. These authors also discuss the possibility of conducting crowding research on offshore oil-drilling platforms. Informal observations and interviews collected on a recent site visit to two platforms differing in density levels are reported. Men on the higher density platform were more reluctant to be interviewed, frequently complained about their living conditions, and were generally unhappy about life on the rig. In contrast, the men on the lower density platforms were quite satisfied with their living conditions, were interested in being interviewed, and had positive feelings about the platform environment and their fellow workers.

The other three chapters of this section present crowding research conducted in university dormitory settings. In Chapter 7, Stokols, Ohlig, and Resnick examine the relationship between college students' perceptions of the crowdedness, physical amenity, and social climate of their residential environments and their sensitivity to crowding in a classroom, their academic performance, and frequency of visits to the campus health center. A series of questionnaires was administered to students at three different time periods during the school year. Perhaps the most interesting finding is that crowding experiences associated with negative perceptions of residential social climate were more clearly associated with behavioral and health problems than those associated with positive feelings toward roommates. The overall pattern of results are discussed in the context of a proposed typology of crowding experiences which posits the experiences in primary environments (where there are frequent social conflicts) will have greater negative consequences than those in secondary environments (where interpersonal conflict is minimized). Additional directions for future field experimental research and the utility of the proposed typology as an environmental design tool are discussed.

In Chapter 8, Holahan and Wilcox take a Lewinian perspective in their report of a field-based evaluation of environmental satisfaction in high-rise and low-rise university residential dorm settings. Residential satisfaction and friendship formation in two types of student residential environments were compared: megadorms housing approximately 3,000 students and low-rise dormitories housing approximately 250 students. More specifically, the study consisted of an analysis of the interaction between students' levels of social competence and type of environment in affecting both residential satisfaction and friendship formation. A survey technique was used that included measures of social competence, satisfaction with the living environment, and the complexity of friendship networks established within the residential setting. The findings indicated that residents of the megadorm were more dissatisfied than residents of the low-rise dormitories with features of the physical environment, student involvement in policy decisions, and the social contacts and support within their residential setting. The implications of studying student adjustment to living environments from an interactionist perspective are discussed.

The final chapter in the first section is concerned with an investigation assessing the utility of a control-based analysis of crowding that we conducted along with one of our students, Lisa Calesnick. The relationship between prolonged exposure to high social density in dormitory settings and motivational deficits characteristic of learned helplessness were examined. Independent samples of "experimental" and "survey" freshmen residents of long- and short-corridor dormitories were tested after 1, 3, and 7 weeks of dormitory residence. The frequency of competitive, cooperative, and withdrawal responses of experimental subjects in a modified prisoner's dilemma situation were gathered along with additional information about subjects' goals during the game. Survey subjects completed a questionnaire that assessed the ways in which they spent their time, their perceptions of crowding, problems associated with dormitory life, and their motivation to assume control over certain situations. Our results indicated that high social density and large residential group size were associated with crowding. Long-corridor residents reported experiencing more crowding, unwanted interactions, less satisfaction, and more problems than did short-corridor residents. More important, however, was the finding that these complaints were associated with reductions in perceived control, which became more salient as length of residence increased. Consistent with Wortman and Brehm's (1975) description of helplessness conditioning, initial recognition of uncontrollable social outcomes aroused negative interpersonal affect and generated attempts to restore control (reactance). With increased exposure to uncontrollable residential conditions, helplessness responding increased.

References

Altman, I. Some perspectives on the study of man–environment phenomena. *Representative Research in Social Psychology*, 1973, 4, 109–126.

Altman, I. Crowding: Historical and contemporary trends in crowding research. In A. Baum & Y. M. Epstein (Eds.), *Human responses to crowding*. Hillsdale, New Jersey: Erlbaum, 1978.

Galle, O. R., Gove, W. R., & McPherson, J. Population density and pathology: What are the relationships for man. *Science*, 1972, *176*, 23–30.

Karlin, R. A., Epstein, Y. M., & Aiello, J. R. A setting specific analysis of crowding. In A. Baum and Y. M. Epstein (Eds.), *Human responses to crowding*. Hillsdale, New Jersey: Erlbaum, 1978.

Wortman, C., & Brehm, J. Responses to uncontrollable outcomes: An integration of reactance theory and the learned helplessness model. In L. Berkowitz (Ed.), *Advances in experimental social psychology* (Vol. 8). New York: Academic Press, 1975.

1

Residential Crowding Research

Paul J. Hopstock, John R. Aiello, and Andrew Baum

As interest in density and crowding has increased, research has considered a growing number of settings and variables associated with crowding stress. Initial investigation of urban density and its relationships with statistical indices of pathology has been supplemented by research using more experimentally oriented approaches, and these studies have considered laboratory settings, grocery and department stores, exterior urban neighborhood spaces, and a number of residential environments. The latter setting has, for a number of reasons, received greater attention than the others and constitutes a major focus of theory and research on human crowding phenomena. This chapter is concerned with the experimental study of residential crowding; we intend to review briefly previous research in this area and to consider the strengths of the experimental and quasi-experimental methodologies that have evolved for use in naturally occurring settings.

Crowding in Residential Settings

Although it is beyond the scope of this chapter to provide detailed summaries of previous research on residential crowding, it is important

Paul J. Hopstock • Department of Psychology, Manhattan College, Riverdale, New York 10471. John R. Aiello • Department of Psychology, Rutgers University, New Brunswick, New Jersey 08903. Andrew Baum • Department of Medical Psychology, Uniformed Services University, School of Medicine, Bethesda, Maryland 20014.

to review the results of this research. One way of doing this is to consider the effects of residential crowding as indexed by the three principal types of dependent measures used. By reviewing cognitive/affective response to residential crowding, stress-related physiological influence, and behavioral consequences of this stress, we can reveal the patterns of response to crowding experienced in residential environments. Where appropriate, we will refer the reader to other chapters in this book for more complete information.

Cognitive and Affective Response

Although there has been some controversy regarding the aversiveness of crowding and density (e.g., Altman, 1975; Freedman, 1975), research on residential crowding clearly suggests that it can be a negative experience. Baron, Mandel, Adams, and Griffen (1976) found that students housed in tripled dormitory rooms (intended for double occupancy) felt more crowded, uncomfortable, and dissatisfied with their room than did students housed two to a room. Tripled students reported their rooms as being more cluttered, full, public, and chaotic than did doubled students, and they seemed to feel less of a sense of control over their living situation. Similarly, Aiello, Epstein, and Karlin (1975) reported that tripled residents felt more crowded than did doubled residents, and they were more negative about their residential experience. (This research will be summarized more fully later in this chapter.)

Additional research has provided evidence of the aversiveness of residential crowding across a number of different situations. Baum and Valins (1977) found that students who felt crowded in their dormitories were also less satisfied with residential experience and expressed more negative feelings about the dormitory and about their neighbors. Bickman, Teger, Gabriele, McLaughlin, Berger, and Sunaday (1973) found that residents of high-rise dormitories rated fellow students as less friendly, independent, secure, thoughtful, warm, and outgoing than did residents of low-rise housing. Further, many of the papers in this volume report evidence of the aversiveness of high-density residential experience. Thus, negative response to residential crowding has been found in descriptions of neighbors, as well as of housing conditions and personal interaction with the environment.

Stress-Related Effects

Exposure to high residential density has also been associated with arousal, stress, and health-related problems. Aiello *et al.* (1975) found that residents of crowded dormitory rooms were more aroused than those that were not crowded, and Baum and Valins (1977) report evidence of stress in crowded dormitory buildings. D'Atri (1975) found heightened levels of systolic and diastolic blood pressure among prison inmates housed under conditions characterized by high social density, and Cox *et al.* (Chapter 6, this volume) report heightened palmar sweat levels for inmates of socially dense prisons. High residential density has also been linked to increasing numbers of medical complaints and visits to health centers (Dean, Pugh, & Gunderson, 1975; McCain, Cox, & Paulus, 1976; Stokols, Ohlig, and Resnick, Chapter 7, this volume).

Behavioral Consequences

A number of studies of behavioral response to crowding have been reported, but studies of behavioral response to residential crowding have not been as numerous. Dean *et al.* (1975) found that naval personnel in high-density settings had more accidents than did those living in less dense settings, and Baum and Valins (1977) found that residents of dormitory environments characterized by high social density performed more poorly when solving anagrams alone than did residents of dormitories housing them in smaller groups. These differences may reflect motivational deficits rather than cognitive difficulties; performance on these tasks appeared to be related to persistence on them, but when residents of socially dense environments were allowed to compete, performance improved markedly. Evidence of motivational deficits and learned helplessness among people living under dense residential conditions has been reported (Baum, Aiello, & Calesnick, 1978; Rodin, 1976; see Chapters 9 and 5, this volume).

Research concerning the effects of residential density on responsiveness to the social environment is also relevant. Subjects experiencing high-density conditions are more likely to withdraw from social interaction with strangers (Baum & Valins, 1977) and to have problems relating to neighbors (McC .rthy & Saegert, Chapter 4, this volume). Residents of high density environments were also less likely to engage in self-disclosure, form group consensus (Baum & Valins, 1977), and to help others (Bickman *et al.*, 1973). The general pattern of these results is that residents of high social density environments engage in withdrawal and decrease their level of social interaction.

These findings have been obtained through a number of means; residents of high- and low-density settings have completed surveys, have been interviewed, have been observed in both naturally occurring and laboratory settings, and have sometimes allowed researchers to make physiological assessments of arousal and stress. As is pointed out in several of the chapters in the volume, laboratory and correlational investigation of residential crowding do not appear sufficient *on their own* to study these settings. The need to assess behavior and experience at a number of different levels, combined with the necessity of maintaining the integrity of the setting under study, has led to the use of quasi-experimental field methodologies. These strategies allow both multilevel assessment and experimental control in the study of response to residential crowding and are generally applicable to most settings. The remainder of this chapter considers the utility of these approaches to the study of residential density and crowding. Research conducted by Aiello *et al.* (1975) at Rutgers University will be examined in detail to illustrate the strengths of this methodology.

Advantages of Field Experimental Techniques

The first step for most experimental or quasi-experimental field studies is to find naturally occurring variations in density among similar subject populations and to study the reactions of residents to those variations. McCarthy and Saegert, for example, studied the reactions of people who were assigned to either high- or low-rise public housing that varied in terms of social density. While some of these studies have considered subjects randomly assigned to residence or have been able to control for differential assignment (e.g., Baron *et al.*, 1976; Baum & Valins, 1977), many have not met this experimental criterion. However, they do provide reasonable evidence of comparability of samples.

There are at least five advantages to this field experimental methodology: (a) subjects are exposed to density conditions for a prolonged period, thus maximizing the likelihood of behavioral consequences; (b) there is a stable subject population, making more feasible longitudinal research; (c) the assessment of cross-situational effects is possible by studying people outside of their residences; (d) causal inferences can be drawn through the use of inferential rather than descriptive statistical techniques; and (e) applicability of findings is assured by studying people in their natural environments.

Prolonged Exposure

Unlike laboratory situations in which people are exposed to density conditions for only a few hours, the use of residential settings allows for the study of long-term reactions to density. Early responses to density conditions often involve behaviors designed to alleviate deleterious effects (Hopstock, 1975); many of the effects of density may therefore only become apparent following the failure of initial coping strategies. Research summarized in this chapter suggests that high density experienced on a long-term basis does generate stress and stress-related symptoms. These effects could not be assessed acurately with laboratory research or through correlation of density and pathology.

In addition, the coping behaviors that people use to deal with high density may themselves incur personal costs for the user. Field experimental studies are uniquely suited to study the effectiveness of various coping strategies, as well as the costs for their use. Baum and Valins (1977) and McCarthy and Saegert (Chapter 4, this volume) have each shown that subjects experiencing social density in residential settings are less responsive to strangers in other situations, and other long-term costs have been noted.

Stable Subject Population

A second advantage of field experiments considering residential settings lies in the stability of the subject population under study. Unlike research in stores or laboratories, experimenters can assume that subjects will change infrequently, and they therefore can be confident of subject availability under constant conditions for later study. These characteristics are essential for the development of a longitudinal research project.

Paulus, Cox, McCain, and Chandler (1975) were the first to report length of residential density effects when they found that tolerance for crowding decreased over time under conditions of high social density. Baum et al. (1978, see Chapter 9) also found effects of length of exposure to high density, noting the development of reactance and helplessness in dormitory settings. These studies used cross-sectional rather than longitudinal techniques, however, and their results could be open to a number of interpretations. The work of Aiello et al. (1975) did use longitudinal techniques and found changes over time in urine cortisol levels, complex task performance, and body orientation during interaction. Research in residential environments using cross-sectional and longitudinal techniques is clearly just beginning, but future research may answer many questions concerning the processes by which people experience the effects of density.

Opportunity to Study Cross-Situational Effects

Another important advantage of field experimental studies of residences is that they allow observation of the effects of density in the environments in which variations occur and in other settings as well. The robustness of density effects can be illustrated by their generalization across a variety of situations. For example, Baum and Valins (1977) found that residents of socially dense dormitories avoided strangers to a greater degree than did residents of less dense dormitories in the dormitory setting *and* in neutral laboratory situations. Stokols *et al.* (Chapter 7, this volume) have also reported evidence of effects that persist beyond the crowded residential setting, and Rodin (1976; see Chapter 5) has found evidence of motivational deficits that generalize to more remote, uncrowded situations. More traditional research strategies appear to be less successful in studying cross-situational effects. Laboratory studies can move subjects from dense to less dense laboratory rooms, but the generalizability of such effects is questionable. Correlational research can provide evidence concerning the reactions of people in a variety of settings but cannot conclusively answer questions concerning the sequencing of variables.

Field experimental studies of residential environments provide one of the few opportunities to study exposure to density that is relatively prolonged and unavoidable for the subject. Under such conditions, experimenters can assume more confidently that generalizability of effect across situations will occur.

Use of Inferential Statistics

The major failing of correlational studies lies in the statistical techniques that they must employ. Scientists interested in the prediction and control of density-related effects seek statistics that allow causal inferences to be drawn from the data they collect. Correlational data cannot be analyzed in that way, and thus many of the questions that scientists ask about the effects of density cannot be answered.

A major advantage that field experimental studies share with laboratory research is the use of inferential statistics. Studies of residential density are somewhat unusual in that density conditions are normally created not by the experimenter but rather by some outside agency or by chance. Thus, experimenters are often limited in the types of density variation available for study. However, as long as reasonable evidence is provided concerning the comparability of sample populations, causal statistics are still appropriate (cf. Campbell & Stanley, 1966).

The ability to draw conclusions about causality are also important if data are to be used in the design or alteration of residential environments. Designers must predict the effects of various architectural changes if they are to create optimal environments in which to live. For example, it is important for them to be able to approximate the optimal ratio of private to semipublic space within buildings so that they can use space effectively and reduce feelings of crowding. Similarly, new design variations should be tested to evaluate their effects. The field experimental approach is appropriate for such assessment and evaluation.

Use of Natural Environments

A fifth advantage of field experimental studies of residential density is their use of natural environments. Considerable discussion of the ecological validity of laboratory research has arisen; Karlin, Epstein, and Aiello (1978) have proposed that the effects of density are situation-specific, and if their proposal is correct, the external validity of laboratory research is limited to the degree to which laboratory experiments simulate or provide analogues for real-world situations. Patterson (1977) has also suggested that most laboratory research is limited in terms of external validity because of setting constraints and the use of college students and other nonrandom population samples.

Field studies of residential settings assure ecological validity by examining the effects of natural environments. The effects of density are most likely to be actualized in such settings because the experience is long-term and realistic in nature and the effects of length of exposure can be determined and specified (e.g., Baum *et al.*, 1978; see Chapter 9). In addition, subjects are normally unaware of the fact that density variations constitute a variable under study, and their responses on dependent variables are less subject to demand characteristics. In such cases, people are not experiencing an unusual event (e.g., a laboratory experiment) but rather are simply adapting to the characteristics of their residence.

The use of natural environments makes field experimental studies highly applicable to real-world problems. Social scientists and designers can work together in natural settings, developing and evaluating various design features that mediate the effects of density. Such interactions between research and design underlie the rationale for most environmental research, and they should be emphasized within the field experimental methodology.

Overview of Field Experimental Research

Settings

Field experimental studies of residential density have been conducted in a wide variety of settings. Perhaps, not surprisingly, the most popular settings for field studies of density have been college dormitories (Baron *et al.*, 1976; Baum & Valins, 1977; Bickman *et al.*, 1973).

College students are readily available as subjects, and college administrative personnel are frequently willing to allow research within their facilities. Jails and prisons have also been studied (D'Atri, 1975; McCain *et al.*, 1976; Paulus *et al.*, 1975). Prisons are particularly suited for density research because they have high physical density ratios and provide residents with little opportunity to avoid the effects of density. Apartment complexes have been considered (McCarthy & Saegert, Chapter 4; Rodin, 1976; see Chapter 5). These settings are particularly important because they closely reflect the residential patterns of an increasingly large segment of the American population. More unusual settings have also been used. Dean *et al.* (1975) have studied the effects of density variables on naval vessels, and Cox *et al.* (Chapter 6) have begun research on offshore oil-drilling platforms.

What all of these settings have in common are density variations, with at least some subjects experiencing high levels of density, and subject populations that are available for study. The use of so many diverse settings has disadvantages and advantages. Results from different settings often present rather different pictures of the effects of density on residents and thus generate some confusion. That confusion is balanced, though, when similar findings in different settings serve to reinforce conclusions concerning the generality of density effects. The breadth of density effects has been well demonstrated by past research, so future researchers might do well to examine in depth the effects of density in specific settings.

Density Variations

Almost as diverse as the number of settings in which residential density studies have been conducted are the number of ways in which density has varied within those settings. Physical density is defined as the ratio of the amount of space available to the number of people occupying that space. There are only two ways to vary density; one can change the amount of space available (spatial density variations) or the number of people in that space (social density variations). However, residential studies make that picture somewhat more complex.

Interestingly, the most commonly used technique for studying density in residential settings has not been to study density variations in private areas, but rather to study architectural effects that promote different social densities in public areas of buildings. McCarthy and Saegert (Chapter 4), Holahan and Wilcox (Chapter 8), and Bickman *et al.* (1973) each studied differences between high-rise and low-rise buildings; Baum and Valins (1977) studied the differences between corridor-design and suite-design dormitories, and D'Atri (1975) compared individual cell versus dormitory units in prisons. The D'Atri study is somewhat unusual because the distinction between private and public areas is somewhat difficult to make in prisons, but its method of determining density was similar to those of the other studies cited.

Variations of social density within housing units were studied by Baron et al. (1976) and Rodin (1976; see Chapter 5). Baron et al. examined the effect of adding a third resident to dormitory rooms that normally house two individuals, while Rodin studied the effects of varying room densities within apartments. Other studies (Dean et al., 1975; McCain et al., 1976; Paulus et al., 1975) have calculated separate indices of social and spatial density and related those indices to a variety of dependent variables.

The diversity of density variations considered in field research has generated little confusion, however. The distinction between social and spatial density effects has been well recognized (Loo, 1977), and most studies have chosen to investigate some form of social density variation. The complexity of density variations merely reflects the complexity of the built environment in which people live.

The Rutgers Study: An Illustrative Example[1]

This chapter will conclude with a summary of research conducted at Rutgers University by the second author, along with Y. Epstein during 1974–1975. This research is distinctive from other field studies in two ways. First, it considered a wide variety of dependent measures of response to crowding. Other researchers (e.g., Cox et al., Chapter 6; Baum & Valins, 1977; McCarthy & Saegert, Chapter 4) have also used a variety of measures but have done so by using different samples for different measures. Thus, the Rutgers research is distinctive in that it was a lon-

[1]This research was initiated in the spring of 1974 after a student of Dr. Aiello's asked appropriately, "If you're so interested in crowding, why don't you do a study of our crowded dorms?" We proceeded to do just that, leaving the security of our lab behind, if only temporarily. During the spring, we interviewed students who were living in the dormitory that we would study to obtain some perspective on the salient dimensions of the person/environment relationships in this setting. We spent much of that summer assembling suitable measures and procedures that would be used in the study. The investigation commenced two days prior to the beginning of fall classes (as soon as students arrived on campus) and terminated at the conclusion of final exams (when students went home for intersession break). The study was supported by grant HD-8546-01 from the National Institute of Child Health and Development to Drs. Epstein and Aiello. Again, we wish to express our gratitude to the many students who assisted us in this investigation. (A more detailed report of this research may be found in Aiello et al., 1975).

gitudinal study of human residential density. The effects of density on individual subjects were assessed a number of times over the course of a semester, allowing conclusions to be drawn concerning the process by which density is experienced by specific individuals.

Setting and Density Variation

The dormitories used in the research were of standard corridor-type design, each bedroom measuring approximately 12 feet by 16 feet. These rooms were designed for two residents, but because of a housing shortage, some of the rooms had to be converted for three-person use. Students were randomly assigned to either doubled or tripled rooms based on a chance lottery, so the experimental requirement of comparability of samples was met. All subjects were freshmen college students from relatively homogeneous backgrounds with respect to socioeconomic status. Thus, the research began with comparable groups of male and female subjects experiencing differing levels of social density.

Dependent Variables

To assess cognitive and affective reactions to density, Aiello and Epstein asked subjects to rate their perceived crowdedness, as well as a large number of items related to their level of satisfaction with living conditions; the stability of room arrangements was also measured. Stress-related physiological and health effects were measured in various ways, two of which will be noted here. First, urine samples were collected from a subsample of subjects twice during the semester and were analyzed for unbound cortisol, which was predicted to increase in the presence of long-term stress. Second, subjects were administered the Cornell Medical Index to measure the number of physical and psychological problems experienced.

The behavioral consequences of density were also assessed in a number of ways. Task performance was measured both early and late in the semester on simple and complex cognitive tasks. The simple task was to cross out the letter *a* every time it appeared in a list of words, while the complex task consisted of a set of nonsense syllogisms whose truth value had to be ascertained by the subject. Responsiveness to the social environment was measured through a self-disclosure questionnaire, self-reported and observed time spent in the room and with roommates, and time invested in personalizing the room. Measures were also taken three times over the course of the semester of variables in actual interaction with roommates (distance, axis, and conversation);

in addition, projected distances of various types of interactions were assessed once at the end of the semester.

Summary of Results

Cognitive and Affective Reactions. Subjects living in tripled rooms reported feeling more crowded and less satisfied with living conditions than did subjects living in doubled rooms. There was also a sex interaction such that the effect of density on both variables was *greater* for women than for men. Aiello *et al*. (1975) hypothesized that males and females used different coping strategies to deal with density, and that the coping strategy employed by males (withdrawal) was more effective than the coping strategy chosen by females (increased levels of interaction). Females, in general, and triple females, in particular, spent more time in their rooms than did the males, suggesting this dualistic approach to coping with the situation.

Data on room stability show similar effects. Tripled rooms were more likely to break up than doubled rooms, though again it was females who showed the most dramatic effects. Thus, social density conditions created by placing a third resident in a room designed for two people apparently did create negative cognitive and affective reactions. It should be noted, of course, that the greater instability of triads than of dyads confounds this social density result.

Physiological and Health Effects. Urine cortisol levels of tripled and doubled residents did not differ significantly, though the observed means were in the predicted direction. Cortisol level of tripled subjects increased, and that of doubled subjects decreased over the course of the semester. Results on the Cornell Medical Index showed that tripled women reported more physical and psychological problems than the other three groups. Thus, the negative consequences of tripling on females, possibly as a function of the greater amount of time they spend in their crowded residential environment, was again demonstrated. These data suggest that long-term density conditions may produce stress-related physiological and psychological problems if adequate coping strategies are not found.

Behavioral Consequences. Task performance on the simple cognitive task improved over the course of the semester for both doubled and tripled subjects. On the complex cognitive task, however, performance differed between groups. Although accuracy of doubled subjects increased, the performance of tripled subjects *decreased* over time.

Tripled and doubled subjects showed few differences in terms of responsiveness to the social environment. Tripled females indicated an initial desire to invest more time than the other three groups in personalizing their room, but this apparently unrealistic intention did not materialize when at the end of 5 months their actual "investment" was appraised. A rather consistent set of density effects did occur on measures of self-disclosure in which tripled women disclosed less to roommates than did doubled women. This effect, however, was reversed for men.

The data on task performance indicated, therefore, that both male and female tripled subjects were affected by the social density variation, though evidence on other measures suggested that females were more influenced by tripling than were males. Future researchers may want to examine more closely the possibility of sex differences in response to residential density.

References

Aiello, J. R., Epstein, Y. M., & Karlin, R. A. *Field experimental research on human crowding.* Paper presented at Eastern Psychological Association, New York, 1975.

Altman, I. *The environment and social behavior.* Belmont, California: Wadsworth, 1975.

Baron, R. M., Mandel, D. R., Adams, C. A., & Griffen, L. M. Effects of social density in university residential environments. *Journal of Personality and Social Psychology*, 1976, *34*, 434–446.

Baum, A., Aiello, J. R., & Calesnick, L. E. Crowding and personal control: Social density and the development of learned helplessness. *Journal of Personality and Social Psychology*, 1978, *36*, 1000–1011.

Baum, A., & Valins, S. *Architecture and social behavior: Psychological studies of social density.* Hillsdale, New Jersey: Erlbaum, 1977.

Bickman, L., Teger, A., Gabriele, T., McLaughlin, C., Berger, M., & Sunaday, E. Dormitory density and helping behavior. *Environment and Behavior*, 1973, *5*, 465–490.

Campbell, D. T., & Stanley, J. C. *Experimental and quasi-experimental designs for research.* Chicago: Rand McNally, 1966.

D'Atri, D. A. Psychophysiological responses to crowding. *Environment and Behavior*, 1975, *7*, 237–252.

Dean, L. M., Pugh, W. M., & Gunderson, E. K. E. Spatial and perceptual components of crowding: Effects on health and satisfaction. *Environment and Behavior*, 1975, *7*, 225–236.

Freedman, J. L. *Crowding and behavior.* New York: Viking, 1975.

Freedman, J. L., Heshka, S., & Levy, A. Population density and pathology: Is there a relationship? *Journal of Experimental Social Psychology*, 1975, *11*, 539–552.

Freedman, J. L., Klevansky, S., & Ehrlich, P. R. The effect of crowding on human task performance. *Journal of Applied Social Psychology*, 1971, *1*, 7–25.

Galle, O. R., Gove, W. R., & McPherson, J. M. Population density and pathology: What are the relationships for man? *Science*, 1972, *176*, 23–30.

Hopstock, P. J. *Experiential differences between social and spatial density.* Paper presented at Eastern Psychological Association, New York, 1975.

Karlin, R. A., Epstein, Y. M., & Aiello, J. R. A setting-specific analysis of crowding. In A.

Baum & Y. M. Epstein (Eds.), *Human response to crowding*. Hillsdale, New Jersey: Erlbaum, 1978.

Loo, C. Beyond the effects of crowding: Situational and individual differences. In D. Stokols (Ed.), *Perspectives on environment and behavior*. New York: Plenum, 1977.

McCain, G., Cox, V. C., & Paulus, P. B. The relationship between illness complaints and degree of crowding in a prison environment. *Environment and Behavior*, 1976, *8*, 283-290.

McGrew, P. L. Social and spatial density effects of spacing behavior in preschool children. *Journal of Child Psychology and Psychiatry*, 1970, *11*, 197-205.

Patterson, A. H. Methodological developments in environment-behavioral research. In D. Stokols (Ed.), *Perspectives on environment and behavior*. New York: Plenum, 1977.

Paulus, P. B., Cox, V. C., McCain, G., & Chandler, J. Some effects of crowding in a prison environment. *Journal of Applied Social Psychology*, 1975, *5*, 86-91.

Rodin, J. Crowding, perceived choice, and response to controllable and uncontrollable outcomes. *Journal of Experimental Social Psychology*, 1976, *12*, 564-578.

Saegert, S., MacKintosh, E., & West, S. Two studies of crowding in urban public places. *Environment and Behavior*, 1975, *7*, 159-184.

2

Crowding and Behavior in Chicago, 1940–1970

Omer R. Galle and Walter R. Gove

That high levels of population density or overcrowding may have severe and negative consequences for human populations appears to be a recurrently intriguing and controversial idea. Although this argument may be found in serious journals of more than 100 years ago (Verhulst, 1845), recently revived interest in the density–behavior equation has been spurred at least, in part, by research on other animal populations such as lemmings, elephants, monkeys, and rats (Calhoun, 1962a) (for a review of the literature on the effect of density on animals see Galle & Gove, 1978). Research on human populations, however, remains fairly sparse and subject to widely varying interpretations (for recent reviews of this literature, see Galle and Gove, 1978; Fischer, Baldassare, & Ofshe, 1975; Carnahan, Gove, & Galle, 1974; Freedman, 1975; Altman, 1975). An earlier article by the present authors (Galle, Gove, & McPherson, 1972) has partly contributed to this continuing controversy. This chapter will attempt to extend the discussion of the question of the relationship between crowding and behavior in human population in two ways. First, after a brief restatement and clarification of the findings reported in our earlier *Science* article, we try to answer some questions that have been raised about that earlier analysis. Second, we examine the consis-

Omer R. Galle • Population Research Center, University of Texas at Austin, Austin, Texas 78705. Walter R. Gove • Department of Sociology, Vanderbilt University, Nashville, Tennessee 37203.
The research for this paper was funded by NICHD grant 2–R01–HD06911–03. An earlier version of this paper was presented at the First World Congress of Environmental Medicine and Biology, Paris, France, 1974.

tency of the findings reported there for Chicago by extending the 1960 analysis to three other time periods, 1940, 1950, and 1970.

Population Density and Pathology in Chicago, 1960: A Reexamination

In April of 1972, *Science* magazine published our article on population density and behavior, which examined data from community areas of Chicago and attempted, at an ecological level, a roughly approximate replication of Calhoun's (1962b) well-known article on population density and pathology in rats. Since the implications of the findings of this earlier article have been questioned by other writers, it seems appropriate to recapitulate briefly how we would state the major findings of that work.

First, we suggested that the situation in human populations is substantially more complex than that for animal populations for several reasons. On the one hand, the fact that man has culture and a complex social structure implies that his social structure is more intricately involved in his behavior, including his pathological[1] behavior. Thus, we constructed indexes of social class and ethnicity in an attempt to take this more complex social structure into account. Based on these indexes, our initial findings suggested that, although there is a significant but relatively small effect of density—measured by persons per square unit of ground area—on rates of pathological behavior, these effects were reduced to insignificance when controls for social structural factors— social class and ethnicity—were introduced. On the other hand, we suggested that the concept of density itself is a more complex factor than one might initially suspect. Our findings suggested that ground-level density, or persons per square unit of ground area, does not necessarily reflect the amount of contact individuals may have at the interpersonal level. To account for this complexity, we decomposed ground-level density into two theoretical components: interpersonal press or crowding, and structural press or density. After this decomposition of density, we reexamined the relationship involving crowding, structural press, social structure, and rates of pathological behavior for large areas of Chicago.

---

[1]The use of the term *pathology* when referring to these rates of behavior is somewhat unfortunate. It was used in our first article because Calhoun used it in his study of rats. The analogy between the "rats" study and the "people" study breaks down to a certain extent in this case, however. We in no way mean to imply, for example, that persons who are divorced or separated from their spouses are in any biological sense more pathological than persons who are single or married (and living with their spouses). We would contend that the major reasons for differences in these rates of behavior in human populations are social-structural and not biological. It may be wiser in future studies to use a less value-laden term to describe these rates of behavior.

The findings of this reexamination have been claimed by both sides of the controversy over whether or not there is a relationship between density and behavior in human populations.[2]

Second, then, on the basis of the ecological[3] analysis of data for Chicago in 1960, we suggested that there *may* be some linkage between high levels of crowding or interpersonal press and high rates of pathological behavior for humans. We would emphasize here the very tentative nature of that statement. Another way of stating this finding is to emphasize one of the contingent observations that accompanies this statement. That is, while we conclude that there may be a relationship between density and behavior on the basis of these data, one other finding emerges much more clearly: density and social structure are so closely intertwined that it is extremely hard, at the ecological level, to separate the independent effects of density as opposed to social structure. To emphasize the high degree of collinearity of the conceptual variables of density and social structure, we present a recomputation of the data in Table 4 of the *Science* article in the first section of Table 1. The first two columns indicate the amount of variance explained (R^2) by the four dimensions of density and the two indexes of social structure. Note that each set of variables does very well at explaining variation in rates of behavior across the 74 community areas of Chicago (admissions to mental hospitals is somewhat of an exception). Note, however, that most of this variance explained is variance that cannot be allocated independently to either density or social structure. Over 90% of the total variance explained by density and social structure combined is variance held in common by density and social structure for mortality, fertility, and juvenile delinquency, while 83% of the variance is common variance for public assistance rates (again, admissions to mental hospitals is somewhat of an exception—only 61% of the total explained variance is common variance). That is, it is explained variance that cannot be attributed either to density independent of the social structural factors or to the social structural factors independent of density. This particular finding of high collinearity seems to have become lost in the ensuing discussions about the relationships between density and behavior. At the same time, however, it is a finding that makes a good deal of sense and one that needs to be reemphasized. After all, it is fairly reasonable that poor people, who cannot afford better housing arrangements, are much

[2]For example, Freedman (1973) cites us as being one of the proponents of the idea that density has severe undesirable effects of human behavior, while at the same time Drop (1972) suggests that her findings from the Netherlands are similar to ours in that virtually no explained variance is attributable to density independent of other factors.
[3]The use of the term *ecological* in this context refers merely to the data being analyzed at an aggregated, areal-unit level rather than at the individual level.

Table 1. Reanalysis of the 1960 Findings on Density and Behavior in Chicago

Pathologies	Variance explained by					
	Density only	Social structure only	Density independent of social structure	Social structure independent of density	Common variance	Total explained variance (R^2)
Recomputation of Table 4 of *Science* article						
Mortality ratio	75.2	68.5	7.1	0.4	68.1	75.6
General fertility rate	73.3	72.8	3.7	3.2	69.6	76.5
Juvenile delinquency rate	84.1	85.0	3.5	5.3	80.6	89.4
Public assistance rate	78.7	78.3	7.4	7.0	71.3	85.7
Rate of admissions to mental hospitals	47.5	29.8	18.1	0.4	29.4	47.9
Reanalysis using *Science* article variables without index construction						
Mortality ratio	79.9	82.5	5.7	8.3	74.2	88.2
General fertility rate	75.9	78.5	4.5	7.1	7.14	83.0
Juvenile delinquency rate	85.5	91.9	1.7	8.1	83.8	93.6
Public assistance rate	86.0	87.4	5.5	6.8	80.5	92.8
Rate of admissions to mental hospitals	49.4	67.3	3.1	21.0	46.3	70.4

more likely to be found living in highly crowded conditions. This statement may appear to be trivially obvious to many, but the strength of this finding over alternatives—density as a major cause of pathological behavior or social structure as a major cause—needs to be stressed. Even though those who wish to shape public policy may desire strong and simplistic statements, these data do not lend themselves to that task. The best we can say at this point is that much more information is needed, information that can be gained only from data more suitable for answering those kinds of questions.

The complexity of the data and their interrelations have also raised a number of questions about the way in which our analysis was executed. We now turn to the task of answering some of these questions.

Questions and Partial Answers: Answering the Critics

The analysis of the Chicago community area data, it has been suggested, had several flaws which might have affected the results that were obtained. Several persons have questioned the validity of the indexes of ethnicity and social class that were used in this analysis (e.g., Ward, 1975). The construction of these measures was admittedly somewhat crude. The weighting factors to add the three individual measures for each index were derived from an examination of the regression coefficients of the three predictor variables (education, occupation, and income) on each of the five pathologies. The weights for constructing the indexes were then developed by examining these equations and attempting to derive weights that would give the highest degree of explained variance in the five pathologies. In this way, we hoped to weight the hypothesis testing in such a way that the social structural variables would explain as much of the variance as they could, and at the same time avoid the problem of having many more variables of one type in the analysis, which tends to inflate artificially the relative amount of explained variance attributable to that set of variables. The second section of Table 1 shows the results of an analysis identical to that in the first section except that the six social structural variables were all entered into the regression equations,[4] i.e., there were no indexes used. Each vari-

---

[4]Note that in Table 1 the total variance explained by density (column 1) ought to be identical for the two sections; they are not. The reasons they are not identical is that a somewhat different procedure was used in the reanalysis. In the original analysis, published in the *Science* article, missing data for several of the pathologies were dealt with by inserting the mean value for all community areas (i.e., all which had that measure reported) for each community area that had a missing value. In the present analysis, community areas that had a missing value were eliminated from that particular regression

able was allowed to contribute independently to the total explained variance. As can be seen in this section, the amount of explained variance that can be independently attributed to the density factors is slightly reduced, whereas the amount of explained variance attributable to social structure is slightly increased. At the same time, however, we would argue that the outstanding feature of the second section is its similarity to the first. That is, the amount of variance explained by either density *or* social structure, independent of the other, is quite small, and the major component is the very high degree of variance explained that cannot be attributed to either density or social structure but is held in common by both general factors.[5]

Several other methodological questions have been raised with regard to the earlier analysis. On the one hand, some have questioned the use of the ratio-type measures of density we used—persons per room, rooms per housing unit, housing units per structure, and structures per acre—because they involve the use of common terms in numerators and denominators, thus inflating or deflating the magnitude of the actual relationship (Schuessler, 1974). A second criticism has been raised because we used measures of central tendency when, in fact, we wish to imply relationships among extreme values (Ward, 1975). That is, we make statements with regard to the relationship involving high levels of crowding, low income families, and juvenile delinquency (for example). As evidence for the relationships among these variables, we present correlations among average persons per room, median family income, and the juvenile delinquency rate, when, in fact, we ought to be employing the percent of all persons living in highly crowded conditions (rather than persons per room) and percent of families with low income (rather than median family income) as predictor variables. In response to these critics, we note, with regard to the ratio-measurement problem, that we

---

analysis. In each case the number of observations was reduced then, but not by much. The biggest reduction was for mortality, in which only 68 (instead of 74) community areas were used in the analysis.

[5]Some have suggested that rather than constructing the indexes of social structure in the manner we did, we ought to have used factor analysis and constructed the indexes from factor score coefficients. In rethinking the analysis, we have explored this alternative. The results of several attempts at constructing indexes led us to reject this method of presenting the data. On the one hand, the social class factors, while showing high intercorrelations, did not load on the factors in a readily interpretable manner. At the same time, it was quite clear that the ethnicity factor was also not readily discernible from the factor loadings. In fact, it seemed clear, both from the preliminary factor analysis and from a close examination of the zero-order correlations with the pathologies, that neither percent foreign-born nor percent Puerto Rican was adding much to explaining variation in the pathologies. The ethnicity factor was in fact primarily percent black (see footnote 22 of the *Science* article).

are not really as concerned with the relations among the ratio variables themselves as with the relationship of the set of ratio measures with the dependent variables. Second, with regard to the use of extreme measures, if the distributions are reasonably normal, it will not greatly affect the nature of the relationships if we use the means (or medians) of the distributions or the percent of the distribution above or below any particular cutting point. If the distributions depart significantly from normality, then the type of analysis we undertake (correlation–regression analysis) would be inappropriate.

Another problem with the measures of density we used (aside from the ratio-measurement problem) is that we had to estimate several of the parts of the ratios—total number of rooms and total number of residential structures—on the basis of open-ended interval data. Because of the incompleteness of the information on these open-ended intervals, as well as the other questions raised about the measures, it is useful to use other measures where they are available. Tables 2 and 3 reexamine the nature of the relationships involving density, social structure, and the rates of pathological behavior in Chicago for 1960 through a different set of variables, designed to take into account the criticisms that were raised about the initial analysis. In Table 2, the zero-order correlations of old measures of each concept are compared with the correlations with each pathology for the new measures.

In the first analysis, we used persons per room as one aspect of crowding or interpersonal press. Its substitute, the percent of housing units that contain households with more than one person per room, has almost exactly the same relationship to each of the dependent variables. The correlations are in every case slightly higher than the ones with persons per room, but the differences are not large enough to be significant. As a substitute for rooms per housing unit, the aspect of interpersonal press that was most closely related to the rate of admissions to mental hospitals, we have used the percent of housing units that are occupied by one-person households. This is, in essence, the opposite end of the interpersonal press or crowding distribution. (It might be more appropriately called isolation.) Here the correlations are somewhat smaller and in the opposite direction from the measure of rooms per housing unit, except for the correlation with the rate of admissions to mental hospitals. Since this is the one pathology most closely related to this aspect of the interpersonal press dimension, it seems an adequate replacement for this measure. (It might be recalled, in fact, that we argue in the original *Science* article that percent living alone is a better measure, from a conceptual point of view, for measuring the relationship between this aspect of interpersonal press than rooms per housing unit.)

For the concept of structural press, we examine two possible surro-

Table 2. Zero-Order Correlations between Pathologies and Selected Variables: Chicago, 1960 Community Areas

Variables	Standardized mortality ratio	General fertility rate	Juvenile delinquency rate	Public assistance rate	Admissions to mental hospitals
Density: Persons/square mile (log)	.376	.362	.492	.361	.349
Interpersonal press					
Persons/room (log)	.825	.812	.784	.891	.398
Rooms/housing unit (log)	-.732	-.604	-.792	-.698	-.684
% in HUs with 1.01+ persons/room	.840	.874	.788	.883	.352
% in HUs that are 1-person units	.576	.407	.422	.327	.509
Structural press					
HUs/structure (log)	.641	.500	.716	.521	.620
Residential structures/square mile (log)	-.374	-.126	-.109	-.138	-.112
% of HUs in structures with 5+ HUs	.576	.431	.674	.583	.630
% of HUs in structures with 10+ HUs	.490	.332	.569	.462	.625
Race-ethnicity					
% of CA population black	.715	.720	.836	.888	.318
% of CA population Puerto Rican	.454	.455	.475	.399	.600
% of CA population foreign-born	-.470	-.520	-.574	-.680	-.108
Education					
Median years of school completed	-.570	-.489	-.437	-.317	-.123
% with less than 8 years schooling	.803	.690	.724	.624	.394
Occupation					
% employed in white-collar occupations	-.565	-.576	-.502	-.459	-.034
% employed in lower blue-collar occupations	.798	.731	.811	.824	.326
Income					
Median family income	-.856	-.838	-.852	-.801	-.497
% families with less than $3,000 income	.933	.844	.863	.921	.559

gates for the measures of housing units per structure and residential structures per acre—the percent of housing units in structures with 5 or more, or 10 or more housing units. Of these two possible substitute measures, it appears that neither is as highly correlated with the pathologies as housing units per structure. The first of the substitutes—percent of housing units in structures with 5 or more housing units—has slightly stronger relationships with the various pathologies, and so we will use this as our substitute measure of structural press.

We have not suggested any new measures for the concept of race and ethnicity. Rather, we suggest using only percent black in the community area as the measure of this dimension. The reasons for this are apparent from Table 2. It is the most strongly related to the pathologies of the three measures. In fact, percent foreign-born is inversely related both to the pathologies and to percent black, and the percent of Puerto Rican heritage is related fairly weakly to the other two measures of ethnicity and to the measures of pathology. For these reasons, we shall exclude both of these measures from the subsequent multiple variable analyses and use only the percent black as our measure of race.

The other three substitute measures of social structure—percent with less than 8 years of school completed for median years of school completed, percent employed in lower, blue-collar occupations (laborers, service workers, and private household workers) for percent employed in white-collar occupations, and percent of families with less than $3,000 annual income for median family income—all seem to be more strongly related at the zero-order level to the rates of "pathological" behavior than the original measures of social structure. This suggests that there may be some truth to the criticism that extreme measures of the social structure variables might change the interpretation of the relationships between density, social structure, and pathology. To answer this question more substantially, we now turn to Table 3.

Table 3 is yet another replication of the original analysis from the *Science* article, this time using the new measures of density and social structure. For density we use two measures. For all pathologies except the rate of admissions to mental hospitals, we use the percent of housing units with one or more persons per room as the measure of interpersonal press, and the percent of housing units in structures with five or more housing units as the measures of structural press. For the rate of admissions to mental hospitals, we substitute the percent living alone for the measure of structural press. We employ four separate measures for social structure: the percent of the community area population that is black, the percent with less than 8 years of school completed, the percent of employed labor force in lower, blue-collar occupations, and the percent of families with less than $3,000 annual income. In our opinion,

Table 3. Density and Behavior in Chicago, 1960: Reanalysis with New Variables for Interpersonal Press, Structural Press, and Social Structure

| | Variance explained by | | | | | |
Pathologies	Density[a] only	Social structure[b] only	Density independent of social structure	Social structure independent of density	Common variance	Total explained variance (R^2)
Standardized mortality ratio	75.8	89.3	0.1	13.6	75.7	89.4
General fertility rate	78.4	73.6	7.4	2.6	7.10	81.0
Juvenile delinquency rate	80.8	82.9	6.3	8.4	74.5	89.2
Public assistance rate	87.1	89.4	1.2	3.5	84.9	90.6
Admissions to mental hospitals[a]	43.6	41.8	17.3	15.5	26.3	59.1

[a]Except for Admissions to mental hospitals, the two density variables are (1) Percent of housing units with 1.01 or more persons per room, and (2) Percent of housing units in structures with five or more housing units. For Admissions to mental hospitals, the second density variable is Percent of housing units occupied by only one person.

[b]The four social structure variables are (1) Percent black, (2) Percent of the population 25 years old or over with less than 8 years of school completed, (3) Percent of families with annual income of less than $3,000 and (4) Percent of the employed labor (male and female) employed in lower, blue-collar occupations (laborers, service workers, and private household workers).

the results in Table 3 once again strongly reinforce our interpretation of our initial results from the *Science* article. That is, there appear to be small but significant portions of explained variance that can be attributed independently both to density and to other aspects of social structure, but the major component of explained variance is that which is held in common by both of these conceptual dimensions. From these results, we may once again say that density *may* have an effect on rates of pathological behavior, but that it is so highly collinear with the other, more traditional measures of social structure (race, education, occupation, and income) that it is very difficult to discern which dimension is most important. That the two measures of density explain on the average 6.5% of the variance (independently of the other four social structure measures) as compared to an average of 8.7% for the four social structure measures is, we feel, less significant than the fact that the amount of variance held in common is, on the average, much higher—66.7%.

Density and Behavior in Chicago, 1940–1970

In the concluding discussion of our first article, we added two other cautionary notes of our own about the interpretation of the findings of that study. First of all, we noted that the data were for only one specific time (cross-sectional) and that they were at the aggregate or ecological level and thus did not necessarily apply to relations at the individual level. In this section of our discussion, we expand that original analysis by looking at similar data for four different times. In a subsequent paper we will be reporting on the effects of crowding indicated in data from a household survey.[6]

Because we have to rely on census data from analysis of the relations between density, social structure, and behavior greatly limits our ability to make strong causal inferences about our results. From a logical point of view the "cause" variables ought to be prior, in time, to the "caused" variables (the effects). Since censuses of population, from which we get the majority of the data, occur only every 10 years, the possibility of making strong causal inferences about the kinds of behavior we are discussing here on the basis of utilizing independent variables from 1950, for example, to predict behavior in 1960 seems very difficult.[7] Ten years is simply too long, given the high rates of inter- and

---

[6]A preliminary draft of this paper was presented in Gove, Hughes, and Galle (1976).
[7]Cross-sectional data may be used in causal analysis *if* the causal system is in "aggregate equilibrium." This means that "either the causal relationships among the variables in-

intracity mobility in the United States, to have any confidence at all that we would be talking about the same set of people at these two different times. One study that does attempt this kind of analysis, however, is somewhat suggestive. McPherson (1973) analyzes data from Chicago for 1950 and 1960. In the two cross-sectional analyses (1950 and 1960), the relationships between density, social structure, and behavior are similar to our own findings. However, he argues that when he attempts to predict 1960 behavior on the basis of 1950 measures of density and social structure, the relationship between crowding and behavior disappears. His conclusion from these findings is somewhat equivocal. He suggests that there still may be a relationship between crowding and behavior, but that the time interval of 10 years is too long a period to measure adequately the necessary lag in the causal sequence. We are inclined to agree. One of McPherson's subsidiary findings raises an additional interesting possibility. He suggests that, at the cross-sectional level of analysis, the relationship between crowding and behavior is stronger in 1950 than it is in 1960. Given the fact that levels of interpersonal press or crowding have greatly decreased in the United States (Carnahan, 1974) (and in Chicago) over the last 30 years, might it not be the case that the relationships between crowding and pathology were stronger in prior, more crowded times but are now decreasing because of the lower levels of crowding? To explore this possibility, and also to try to confirm (or refute) our initial 1960 findings with different sets of observations at other times, we collected information on density, social structure, and rates of pathological behavior for community areas in Chicago for 1940, 1950, and 1970 to add to our 1960 analysis. We have collected two other additional sets of data. First, we have obtained rates of juvenile delinquency for years other than the census years, which will allow us to test, for this one "pathology," the suggestion from McPherson that the appropriate time lag between "cause" (density and social structure) and "effect" (juvenile delinquency) is less than 10 years.[8] Second, we have also collected similar information on a sample of 80 census tracts (much smaller areal units than community areas) in Chicago for 1970. These 80 tracts were selected purposively, rather than randomly, in an attempt to maximize the independent variation in levels of crowding and social structure. (The sample of tracts is discussed in more detail in the Gove *et*

volve simultaneous causation, or that the changes in the independent variables have had time enough for the consequent changes in the dependent variables to occur" (McPherson, 1973, pp. 92–93).

[8] See Shaw and McKay (1969). In addition, we received the juvenile delinquency rates for the 1967–1970 period from Mr. McKay at the Institute for Juvenile Research in Chicago. The "lag" period varies for each decade. For the 1940 lag period, rates for 1945–1951 were used; for 1950, rates for 1954–1957 were used; for 1960, rates for 1962–1967 were used.

al., 1976, paper.) Thus we will look at five different sets of data at four different times.

For this over-time analysis, we shall focus on only four measures of pathological behavior—fertility, mortality, juvenile delinquency, and marital instability. For the first three measures, we were able to obtain general fertility rates, age-adjusted mortality rates,[9] and juvenile delinquency rates for 1940, 1950, 1960, and 1970 for community areas. Unfortunately, these rates are collected only at the community area level and not for census tracts. We therefore were forced to assign to each of the census tracts the mortality, fertility, and juvenile delinquency rate for the community area in which the census tract was located. Since several tracts from the same community area were included, and several community areas were not represented in the tract sample, the variance on these pathology rates is considerably reduced. However, we were able to obtain a measure of marital instability for the census tract sample and for 1940 and 1960 community areas: the percent of the adult population that were separated or divorced.[10] We do not include either the rate of admission to mental hospitals or the rate of public assistance to persons under 18 years of age, as we were able to obtain these measures only for 1960.[11]

For our measures of density and social structure, we also attempt to employ a consistent set of measures. For interpersonal press or crowding, we used the percent of housing units that have 1.01 or more persons per room. For structural press, we continue to use the percent of hous-

---

[9]Our age-adjusted mortality rate is the crude death rate divided by the percent of the population over 65 years of age. In 1960 we did have a standardized mortality ratio and in 1950 a (directly) standardized death rate. There are, it is admitted, better age-adjusted mortality rates than the one we use. The one we use, however, is the only one we were able to obtain for all four time periods.

[10]In 1950, the percent separated was not reported as a separate category, and the percent widowed was included with percent divorced. In 1970, there was no *Local Community Fact Book* produced by the Chicago Community Inventory. As a consequence, the percent separated and divorced simply was not readily available for community areas, although it was available for census tracts.

[11]If overcrowding poses a serious problem, we would expect that individuals living in highly crowded circumstances would tend to be in poor mental health. In fact, it seems likely that poor mental health may, to some extent, act as an intervening variable between overcrowding and some forms of social pathology. For example, juvenile delinquency may, in part, be a reaction to the poor mental health of the parents, which in turn is due to overcrowding. Our measure of poor mental health, admissions to mental hospitals, may not, however, be a good indicator of the reaction to overcrowding. Persons are typically admitted to mental hospitals for a very serious disorder, one that frequently has genetic components and often reflects a history of very seriously impaired social relationships. It is thus highly plausible that overcrowding may produce poor mental health but not mental illness (i.e., a disorder requiring hospitalization).

ing units in structures with five or more housing units. Similarly, we use for our measure of race the percent black, and for our measure of occupation the percent of the employed labor force in lower, blue-collar occupations. For education, we use the percent with less than 8 years of school completed in 1950, 1960, and 1970. In 1940, since we cannot get this measure, [12] we use the percent with less than 5 years of school completed. Because of the dramatic changes in income levels, the measure of low income changes for each of the time periods: in 1940 it is the percent of individuals with earnings of less than $1,000; for 1950, it is the percent of families with less than $2,000 annual income; for 1960, the percent of families with less than $3,000 annual income; for 1970, the percent of families below the poverty line as defined by the census.

Table 4 presents the result of the multiple regression analysis for the two measures of density and the four measures of social structure on each pathology that we were able to obtain. Again, the overall results appear to be consistent with our initial interpretation of the 1960 results. Looking first at community area level analysis for all time periods and all pathologies available, the two density variables are able to account for an average of 6.6% of the explained variance independently of social structure. Social structure is able to account for slightly more—12.3%—of the explained variance independent of density factors. The amount of explained variance held in common by the two sets of variables is somewhat smaller for the entire set than for 1960—only about 50% (57.7% is the average). The analysis of the sample of 1970 census tracts shows slightly different results. A lower amount of total explained variance (an average of 56.0%) and common variance (an average of 22.9%) appears to be the case here, along with higher levels of explained variance attributable to density factors alone (12.8%) and to the social structural variables (20.3%). The reasons for this are not entirely clear. On the one hand, although we expected less variation on the dependent variables, given the fact that a number of census tracts had the same values for the measures of mortality, fertility, and juvenile delinquency, the standard deviations for all variables—dependent as well as independent—in the census tract sample were higher than their counterparts in the community area level analysis. Apparently, the selection of census tracts in a manner that minimized the collinearity between the social structure and density allowed for more extreme variations on all the variables, whereas, as would be expected, the effects of density and the social structure show more from independent effect.

[12] In 1940, years of school completed were reported for the following categories: (1) no school, (2) 1–4 years of school completed, (3) 5–8 years of school completed. In other years, that third category was 5–7 years of school completed, and 8 years of school completed was reported for a separate category.

Table 4. Partitioning of Variance Explained for Each Pathology by Density and Social Structure: Chicago, 1940–1970

Pathologies	Variance explained by					
	Density only	Social structure only	Density independent of social structure	Social structure independent of density	Common variance	Total explained variance (R^2)
Fertility						
CAs 1940	5.8	9.8	0.8	4.8	5.0	10.6
1950	46.0	29.5	19.9	3.4	26.1	49.4
1960	78.4	73.6	7.4	2.6	71.0	81.0
1970	61.0	81.9	2.7	23.6	58.3	84.6
CTs 1970	40.0	55.1	3.8	18.9	36.2	58.9
Mortality						
CAs 1940	54.7	55.9	4.9	6.1	49.8	60.8
1950	82.5	67.3	15.9	0.7	66.6	83.2
1960	86.5	75.7	11.3	0.5	75.2	87.0
1970	58.2	72.4	4.4	18.6	53.8	76.8
CTs 1970	19.4	42.8	1.3	24.7	18.1	44.1
Juvenile delinquency						
CAs 1940	63.8	80.9	2.3	19.4	61.5	83.2
1940 lagged	60.3	75.4	4.3	19.4	56.0	79.7
1950	65.8	88.7	1.2	24.1	64.6	89.9
1950 lagged	55.8	72.1	0.2	16.5	55.6	72.3
1960	80.8	82.9	6.3	8.4	74.5	89.2
1960 lagged	73.3	73.7	9.2	9.6	64.1	82.9
1970	61.5	77.7	5.6	21.6	55.9	83.1
CTs 1970	46.3	33.6	23.5	10.8	22.8	57.1
Marital instability						
CAs 1940	70.3	80.7	9.1	19.5	61.2	89.8
1960	88.5	91.5	6.6	9.6	81.9	98.1
CTs 1970	36.9	41.4	22.4	26.9	14.5	63.8

The three specific times during which we were able to try a "lagged" analysis on juvenile delinquency showed rather mixed results. The 1940 variables seem to show a slightly higher degree of explained variance attributable to the density factors when predicting juvenile delinquency rates for the 1945–1951 period than the rates for the period around 1940. A similar pattern appears for using 1960 census measures to predict juvenile delinquency rates for 1962–1965. At the same time, however, the results for 1950 show virtually no explained variance attributable to density factors alone for either the 1950 period or the 1954–1957 rates of juvenile delinquency. In none of these cases do the differences between the cross-sectional and lagged analyses appear to be large enough to justify any strong statement about the relative efficacy of one type of analysis over the other. Overall, the results are consistent with the general pattern for all variables and time periods.

In summary, then, our several analyses of data at the ecological level for Chicago reconfirm our interpretation of the findings of our *Science* article. That is, although social structural factors are highly intercorrelated with high levels of structural and interpersonal press, there is a small but significant amount of explained variance that can be attributed to density independently of other social structural factors. This relationship holds for several different time periods for Chicago, as well as for at least two different types of areal units.

References

Altman, I. *The environment and social behavior*. Monterey, California: Brooks/Cole, 1975.

Calhoun, J. B. The ecology and sociology of the norway rat. Public Health Service Publication No. 1008. Washington, D.C.: U.S. Department of Health, Education and Welfare, 1962. (a)

Calhoun, J. B. Population density and social pathology. *Scientific American*, 1962, 206 (2), 139–148. (b)

Carnahan, D. L., Gove, W. R., & Galle, O. R. Urbanization, population density and overcrowding. *Social Forces*, 1974, 53(September), 62–72.

Deevey, E. S. The hare and the haruspex: A cautionary tale. *Yale Review*, 1960, 49, 161–179.

Drop, M. Population density and well-being in Dutch towns and cities. Unpublished manuscript, Nederlands Institut voor Preventieve Geneeskunde Tno, 1972.

Fischer, C., Baldassare, M., & Ofshe, R. Crowding studies and urban life: A critical review. *Journal of the American Institute of Planning*, 1975, (November), 400–418.

Freedman, J. *Population density and pathology: Is there a relationship?* Paper presented at a meeting of the American Psychological Association, Montreal, Canada, August 29, 1973.

Freedman, J. *Crowding and behavior*. New York: Viking, 1975.

Galle, O. R., & Gove, W. R. Overcrowding, isolation and human behavior: Exploring the extremes in population distribution. In K. Tauber, L. L. Bumpass, & J. A. Sweet (Eds.), *Social demography*. New York: Academic Press, 1978, pp. 95–132.

Galle, O. R., Gove, W. R., & McPherson, J. M. Population density and pathology: What are the relationships for man? *Science*, 1972, *176*(April), 23–30.

Gove, W. R., Hughes, M., & Galle, O. R. *Overcrowding in the home: An empirical investigation of its possible pathological consequences.* Presented at the meeting of the American Public Health Association, 1976.

Laws, R., & Parker, I. *Symposium of the Zoological Society* (Vol. 21). London: Academic Press, 1968, pp. 319–359.

McPherson, J. *A question of causality, a study in the application of regression techniques to sociological analysis.* Doctoral dissertation, Vanderbilt University, 1973.

Schuessler, K. Analysis of ratio variables: Opportunities and pitfalls. *American Journal of Sociology*, 1974, *80* (2), 379–396.

Shaw, C. R., & McKay, H. D. *Juvenile delinquency and urban areas.* Chicago: University of Chicago Press, 1969.

Susiyama, Y. Social organization of hunaman langurs. In S. A. Altmann (Ed.), *Social communication among primates.* Chicago: University of Chicago Press, 1967, pp. 221–236.

Verhulst, P. E. *Nouveaux mémoires de l'Academie R. des Sciences de Bruxelles*, 1845, *18*, 1.

Ward, S. K. Methodological considerations in the study of population density and social pathology. *Human Ecology*, 1975, *3*(4), 275–286.

3

Crowding in Urban Environments: An Integration of Theory and Research

Donald E. Schmidt

With the growing concern over the quality of life in our large urban centers, the topic of human crowding has taken on importance as a contemporary research problem. The observable conditions of over-population and the recognized trend toward increasing urbanization on a worldwide basis (Davis, 1965) have led social scientists and urban designers alike to pose questions about the impact of density on the lives of city residents. Despite the rapidly increasing body of literature dealing with this area, however, primary questions still remain unanswered, namely: What is the relationship between population density and the individual's perception of crowding in his living environment, and what are the psychological and physiological consequences of being crowded on a long-term basis?

In recent years, the popular press has attempted to draw analogies between the aberrant behavioral patterns and biological abnormalities found in animal studies and the social and medical problems typically noted in cities. The validity of such a comparison is tenuous at best and does not give us an accurate picture of what is taking place in human populations. Certainly, higher levels of phylogenetic development have increased man's ability to adapt to his external environment (Glass & Singer, 1972; Schneirla, 1971). Additionally, most laboratory studies that have investigated density and crowding fail to speak directly to condi-

Donald E. Schmidt • Societal Analysis Department, General Motors Research Laboratories, Warren, Michigan 48090. This research was supported, in part, by National Science Foundation Grant GY–11516.

tions of urban living and remain to be interpretated in this expanded context.

The study of crowding in human populations is a complex topic. It is clear that the relationship between physical measures of population density and the individual's subjective experience of being crowded is far from perfect (Stokols, 1972a, 1972b). The importance of both situational conditions and individual differences in cognitive evaluations have become increasingly apparent (Buttimer, 1972; Carr, 1967; Freedman, Levy, Buchanan, & Price, 1972; Hall, 1966; Lee, 1966; Schmidt, Goldman, & Feimer, 1976).

Theoretical approaches that treat individual and situational differences as a central concern can be subsumed under two primary category headings. The first of these approaches postulates that the perception of crowding is inversely related to the individual's ability to exercise behavioral freedom and control over the social and physical environment. For the most part, this position is modeled after Brehm's (1966) theory of psychological reactance. Brehm hypothesized that maintenance of freedom of choice is an important concern strongly related to an individual's affective and behavioral responses. He argued that people are disposed to maintain or restore freedom when it is threatened, and that an individual's reaction to the environment is dependent on his success at accomplishing this end. Proshansky, Ittelson, and Rivlin (1970) applied this theory to crowding. They postulated that crowding could be situationally defined as a condition in which the number of people present were sufficiently large to reduce an individual's behavioral freedom and choice. Esser (1973) defined crowding as the inability to complete the man–environment transaction in which one is engaged. This occurs when social or physical factors block the individual from dealing effectively with a given situation. Similarly, Stokols (1972b) has noted that a person will feel crowded when the demand for space required by a specific activity exceeds the available supply.

This theoretical position has been given further support by the experimental results of Glass and Singer (1972) who found that perceived control over the onset, offset, or predictability of an environmental stressor (noise) effectively reduced the occurrence of negative aftereffects. Sherrod (1974) partially replicated this result using density as the environmental stressor.

The ecological approach to environmental problems offers an explanation that is similar to the freedom and control position. The embodiment of Wicker's (1973) research on overmanned situations is the conceptualization of crowding in terms of behavioral restriction, competition for scarce resources, and social interference by others present in a setting (Saegert, 1973; Stokols, 1976).

The second theoretical approach to crowding deals with the effects and consequences of perceptual and cognitive overstimulation. Wohlwill (1966) has described the investigation of environmental perception and related areas as a study in the psychology of stimulation. Milgram (1970) has discussed the concept of "cognitive overload" as a condition that is related to stress and social withdrawal in cities. Urban congestion and visual complexity overwhelm the individual's perceptual capacities. In a similar vein, Desor (1972) has conceptualized the perception of crowding as a cognitive response to excessive social stimulation. Finally, Kutner (1973) has suggested that crowding may be related to the inability of the individual to shield him- or herself from the impingement of external social stimulation. The person is unable to avoid unwanted social or visual interference (Altman, 1975). The stimulus overload position, similar to the behavioral constraint explanation, is also partially concerned with a loss of control over the environment. Urban residents are unable to reduce excessive amounts of social contacts or sensory overstimulation (cf. Altman, 1975; Saegert, 1973; Valins & Baum, 1974). The stress resulting from this external condition is related to the experience of crowding.

Crowding has also been described as an attributional process. That is, the perception of crowding may involve an evaluative process based on the individual's attempt to attribute general arousal to a specific source in the environment (Gochman, 1976; Worchel & Teddlie, 1976). Thus, salient conditions such as density may be blamed for the individual's negative affective reaction to a specific setting (see Freedman, 1975; Schachter & Singer, 1962).

Perhaps the common thread that runs through both the freedom and control and the overstimulation approaches is the underlying cognitive basis of the perception of crowding. This cognitive explanation of crowding would imply that physical and social conditions found in urban environments will be perceived and evaluated selectively by each individual, and that crowding is best viewed as a complex interaction between the man and his external environment (Carr, 1967; Esser, 1973; Lee, 1966; Sonnenfeld, 1966; Schmidt, Goldman, & Feimer, 1976; Stokols, 1976). Mediation by organismic factors has emerged as a central concern in the research on human crowding. For example, Stokols, Rall, Pinner, and Schopler (1973) noted that personal and social factors may act to sensitize the individual to actual or potential constraints in a setting. This increased sensitivity may increase the likelihood that an individual will experience crowding. Carr (1967) suggested that the selective perception and evaluation of the environment is based on the individual's past experiences. In this light, the evaluation of crowding is somewhat dependent on the values, attitudes, and expectations that are held

by the urban resident. Thus, the perception of crowding is largely determined by the evaluative standards of the individual and by the behaviors, activities, and settings involved (Carr, 1967: Freedman *et al.*,
1972; Hall, 1966; Lee, 1966; Saarinen, 1969; Schmidt *et al.*, 1976; Stokols,
1972b, 1976).

The urban environment represents a complex array of both physical
and social factors that must be taken into account in any thorough investigation of the parameters that potentially affect the perceptions of the
city resident. Each individual will be exposed to a number of different
settings in his or her normal day-to-day activities. In addition, each of
these different areas will vary in its social-environmental characteristics,
not only across settings but on a temporal basis as well (Buttimer, 1972;
Schmidt *et al.*, 1979). It would seem important to test the theoretical
positions that have been discussed previously within a general urban
environment. It is also of interest to investigate each urban level (i.e.,
residence, neighborhood, and city) separately since different factors
may be associates of the perception of crowding at each of these distinctive spheres of social interaction and activity.

A Field Study of Urban Crowding

The ambiguity of the findings from the initial animal studies and the
limitations of the microenvironments typically used in experimental
studies of crowding made it apparent that much of the difficulty in the
accurate conceptualization of the perception of crowding might be al-
leviated by a field study directed to test the physical, social, and personal parameters of this evaluation. The location that was selected for
this study was San Bernardino–Riverside, California, a two-community
area located approximately 60 miles southeast of the large Los Angeles
metropolitan basin. This area was of special interest since it was in active
transition from a rural-suburban area to one distinctively urban in nature.

The method of study involved a relatively large-scale in-community
questionnaire/survey technique. An 83-item questionnaire was developed to measure a global range of social-environmental conditions,
personal characteristics, and attitudes toward overpopulation and development in the city. In addition to these variables, a number of
physical indices were collected that measured the physical density of the
residence and the census tract in which the residence was located, as
well as the distance measurements of the residence from a number of
land-use amenities (i.e., parks, major roads and freeways, commercial
or industrial areas, and open-space zones). The perceptions of crowding
in the residence, neighborhood, and city were also gathered.

Sampling was conducted in a way that allowed us to obtain a representative distribution across a number of demographic and physical variables. Sampling areas were chosen to include all constituent ethnic and income groups. In addition, sampling times were scheduled so that a representative survey of both working and nonworking residents could be obtained. Finally, census data books were used in conjunction with aerial photographic slides of the two cities, allowing us to select sampling census blocks that offered a diverse variety of physical characteristics in regard to the location of residences. A total of 698 questionnaires were collected.

Theoretical consideration and a factor analysis of the questionnaire item responses were used to form 16 composite scales that conformed to minimal reliability criteria and measured important aspects of the environment, personal attributes, and attitudes. A description of these scales and their respective reliabilities is reported in Table 1.

The relationship between the psychological and physical factors and the perception of crowding at the various levels were tested using a multiple linear regression analysis. The perception of crowding responses for the residence, neighborhood, and city were used, in turn, as criterion measures, and the physical measures and psychological scales were used as predictors. The variance accounted for by each of these levels of analysis and predictor set is detailed in Table 2. Since ethnic and income-related factors might have confounded the results (i.e., the quality of the living environment may be largely a function of one's income and/or ethnic classification and hence have been the major influence on the results), the effects of these demographic variables were regressed out of the analysis. This was accomplished by entering the ethnic or income variables into the regression equations on the first step and the psychological and/or physical predictors on the second. The resulting variance values represent the percentage of the variation in the perception of crowding that can be explained uniquely by the physical and psychological factors, independent of income or ethnic classification. The statistical techniques employed can be found in Kerlinger and Pedhazur (1973).

Analysis at the Residential Level. At the residential level of analysis, physical and psychological predictors proved to be equally important. The primary physical measure was residential density. Three psychological variables, indicative of both density-related factors and personal attributes, also contributed to the prediction of the perception of crowding at this level. These were the respondent's ability to attain the desired level of privacy, the freedom to get away from the residence, and the importance of spatial factors in the selection of the current residence.

Table 1. Psychological Scales Description and Reliability

Scale	Description	Reliability
1. Perception of crowding in the residence	Respondent's nonrelative perception of crowding in the residence	.787
2. Aesthetics and order in the city	Subjective perception of the city in terms of beauty and order	.627
3. Freedom to get away	Ability of the respondent to get away from the house and city when desired	.770
4. Crowding in shopping areas	Respondent's subjective perception of crowding in shopping areas in the city	.763
5. Perceived change in neighborhood density	Perceived change in neighborhood density levels since moving into residence	.778
6. Crowding in public recreational facilities	Perceived crowding in public recreational areas and how it effects personal enjoyment	.717
7. Association with neighbors	Self-report of amount and frequency of interaction with neighbors	.617
8. Attainment of privacy	Self-report of ability to attain desired privacy	.680
9. Attitudes toward limitation of development in the city	Attitudes toward further development in the neighborhood and undeveloped areas of the city	.777
10. Preference for conditions related to population in the city	Preferences related to conditions of crowding and relative number of people living in the city	.692
11. Perceived homogeneity/ heterogeneity of neighbors	Respondent's perception of neighbors as having similar or different attitudes, beliefs, and backgrounds relative to self	.632
12. Spatial factors in choice of present residence	Respondent's self-report of the importance of house and yard space in choice of residence; also importance of less crowded conditions in specific section of town	.789
13. City compared to other cities relative to crowding	Respondent's perception of crowding in city compared to other Southern Californian cities	.688
14. Solutions to the effects of overpopulation	Attitudes toward proposed solutions to overpopulation: more open space, development of additional recreational areas, effective planning by government	.620
15. Problems related to traffic in the city	Respondent's reaction and perception of traffic and congestion on city thoroughfares	.593
16. Attitudes concerning overpopulation as a related factor of social problems	Overpopulation related to violent crimes, mental illness, medical pathologies, pollution, and social alienation	.936

Table 2. Squared Multiple Correlation $(R^2)^a$ by Type of Predictor and Level of Crowding Controlling for Ethnic Grouping and Income

	Level of analysis		
Type of predictor	Residence	Neighborhood	City
Psychological scales			
Initial prediction	22.1	31.0	39.3
Controlling for			
Ethnic grouping	19.7	29.7	37.1
Income	20.4	28.6	38.4
Physical measures			
Initial prediction	22.0	5.0	0.0
Controlling for			
Ethnic grouping	16.0	4.3	—
Income	19.8	3.0	—
Both psychological scales and physical measures			
Initial prediction	32.3	33.6	39.3
Controlling for			
Ethnic grouping	26.5	32.9	37.1
Income	30.0	30.8	38.4

a R^2 value corrected for shrinkage to provide unbiased estimate. Tabled value represents the percentage of variance accounted for by the specific level of analysis and predictor set.

Analysis at the Neighborhood Level. The importance of physical measures at the neighborhood level of analysis declined markedly from that indicated in the residential analysis. Perhaps this is due to the lack of an adequate neighborhood density index. However, the lack of a correlation between the perception of crowding in the neighborhood and the census tract density measure, as well as wide response diversity observed from one contiguous residence to the next, would confirm the reduced importance of density. Although density measures were uncorrelated with the perception of crowding in the neighborhood, the distance measurements of the residence from the land-use amenities added a weak but significant contribution to the prediction of crowding at this level. The psychological predictors in this analysis consisted primarily of personal evaluative factors. Preferences for population-related conditions in the city, problems related to traffic, spatial factors in the selection of the residence, perceived changes in neighborhood density, and crowding in shopping areas were all significantly associated with the perception of crowding. In addition, the privacy factor also proved to be an important predictor.

Table 3. Predictors of Crowding at Three Levels of Analysis

Step	Residence	Neighborhood	City
1	Density in house (people/room)	Preferences for conditions in city	Preferences for conditions in city
2	Attainment of privacy	Problems related to traffic	Problems related to traffic
3	Lot density (people/lot size)	Perceived change in neighborhood density	Crowding in recreational facilities
4	Freedom to get away	Attainment of privacy	City compared to other cities relative to crowding
5	Spatial factors in choice of residence	Spatial factors in choice of residence	Attitudes toward limitation of development
6		Crowding in shopping areas	Attainment of privacy
7		Distance from parks	
8		Distance from commercial/ industrial areas	
9		Distance from major roads	

Analysis at the City Level.   The decrease in the importance of the physical densities and measurements is also apparent at the city level of analysis. There were no predictors of this type that were significantly related to the perception of crowding at this level. Three types of psychological predictors were important. The first of these are related to personal evaluative factors. The preference for population-related conditions, attitudes toward development in the city, and the ability to attain desired privacy were all included in the regression equations. The second type of psychological predictors is indicative of individual perceptions relative to conditions in the city. Problems related to traffic and crowding in public recreational facilities added significantly to the prediction of the city crowding variable. Finally, crowding in the city compared to other Southern California cities was indicative of general comparative evaluations that were also correlated with the perception of crowding. A summary of the regression analyses and the correct predictor ordering are presented in Table 3.

Discussion and Implications

Perhaps one of the most interesting findings of this study was the inverse relationship between the level of crowding analysis and the importance of physical density measures. Residential density was the most potent predictor of the perception of crowding in the residence. In addition, two of its psychological concommitants, privacy and the importance of spatial factors, were also included. Mitchell (1971) found that density of the residence was negatively related to self-reported privacy. The current study has replicated and extended this finding.

The initial importance of spatial factors in the selection of the residence is positively related to the perception of crowding. This predictor scale may be more important, however, in its mediating interaction with residential density as a determinant of the perception of crowding. For example, if spatial factors were initially important to the resident, then he/she might be more sensitive to the restriction of space. For these people, perceived spatial restriction in the residence might be directly translated into the inability of the individual to gain the desired degree of privacy. Density would be more salient and these people would be more likely to experience negative affect. Buttimer (1972) noted that individual and group-related differences in sociospatial reference systems, the evaluative standards and expectations relative to the spatial requirements of an activity, may be important when designing residential areas. Individual sensitivities could be primary determinants of perceived limitations (Carr, 1967; Stokols et al., 1973). The freedom and

control approach would allow that each person applies different criteria
to these evaluative judgments. In the residential analysis, this point is
accentuated by the nature of the psychological predictors. It appears that
when the individual is unable to attain the desired degree of privacy, is
concerned with spatial factors, and perceives a lack of control over "escape" from the situation, the perception of crowding is maximal.

Similarly, factors that aligned with the perception of crowding at
the neighborhood and city levels of analysis are less immediate, and
presumably more avoidable. Their effects upon the urban resident are
less dependent on associated physical conditions, and more related to
situational and personal variables. Specific levels of density have no
systematic relationship to the perception of crowding. However, density
as it functionally affects behaviors and activities may be an important
factor. Hence, crowding is not dependent directly on density but rather
on its ecological (distributional) characteristics and concommitant psychological impacts (Rapoport, 1975).

The freedom and control theoretical approach would predict that
the perception of crowding is not strictly contingent on physical measures unless they exert direct limitations on behavioral freedom, choice,
and control for the individual. This logic may be especially valid at the
city level of analysis. While we may initially predict that the perception
of crowding at the city level would be determined by a cognitive composite of crowding in a number of component areas, this finding did not
occur. For instance, although problems related to traffic was an important predictor, crowding in public recreational facilities was a weak predictor, and crowding in shopping areas was uncorrelated with the dependent measure. Second, the physical census tract density measure
was also uncorrelated with the crowding variable. Based on this result,
we may hypothesize that periods of crowding in shopping and recreational
areas are both predictable and avoidable, and that the urban resident's
ability to predict these conditions and/or to exercise control over interactions in these settings may effectively mitigate the effects of high levels
of physical density. For example, we can readily determine that shopping centers and recreational facilities will reach maximum density
levels during weekend periods. This predictability would allow the city
dweller either to choose to avoid these areas in some cases or to generate
accurate expectancies prior to venturing into them.

The overstimulation definition of crowding received only scant
support in these data. The privacy factor may be interpreted in this
context. In this sense, the lack of privacy would be defined as the inability of the individual to avoid social contacts with others in his/her
living and interaction settings (Altman, 1975; Desor, 1972; Kutner, 1973;
Saegert, 1973). Certainly, the appearance of the privacy predictor at all

three levels of the analysis would offer support for this hypothesis. However, the aesthetics and order predictor scale did not provide us with significant information at any level of analysis. This finding would weaken the cognitive overload hypothesis. In another sense, however, privacy may be viewed as a factor that is indicative of a lack of control on the part of the individual resident. That is, limitations on one's ability to attain privacy may be perceived as a restriction of control over interactions with others rather than as the product of sensory overstimulation (Altman, 1975). Certainly, these theoretical positions are not mutually exclusive viewpoints and both interpretations may be accommodated.

The implications of these findings for urban design are apparent. The design of residential, recreational, and commercial areas in the city must implement planning options that will allow the individual to exercise more control in his/her living environments. Similarly, artificially applied architectural manipulations may increase the *perception of space*, as well as freedom and control when actual spatial limitations are unavoidable. The attributional hypothesis of crowding would suggest that simple design techniques that reduce the salience of too many people, and focus attention on other aspects of the environment may reduce the perception of being crowded (Worchel & Teddlie, 1976).

It has been implied that the state of the urban environment may inherently allow the resident certain degrees of control and predictability. However, the effectiveness of this predictability and control is limited. For example, general cultural behavioral patterns may limit the individual from exercising any great degree of control over his/her environment. The very predictability of the less immediate areas in the neighborhood and city is at least partly based on an individual's knowledge about social ecological patterns in the city. However, the existence of such patterns may, in turn, preclude the individual from exercising control over the environment. For example, while many areas in the city reach minimum density levels during the normal eight-to-five workday, many individuals are unable to take advantage of these uncrowded conditions. Similarly, when density levels in areas of the city are maximized during weekends, many residents are forced, by practical considerations, to enter these settings. High density in many city areas is tantamount to behavioral limitation and restriction. Thus, the temporally based ecology of cities may impose pragmatic limitations upon urban residents. Of course, it is not within the realm of the urban designer to change general societal behavioral patterns. Nevertheless, design options that act to more evenly disperse the population during these peak density periods, or that reduce the salience of high density, may help reduce perceived interference and its accompanying negative affect.

The indication that crowding is largely a cognitive phenomenon could complicate any simple efforts to deal with it. Of more importance, however, is the diversity of conditions that are apparent in different cities. Certainly, if the causes of crowding within any given geographic area are to be effectively dealt with, the urban designer will have to involve himself in a dialogue with the residents of these areas to find out which factors are adversely effecting different portions of the general population. While the restriction of behavioral freedom and control has been shown to be a viable definition of crowding, it is clear that the exact manifestations of this theoretical explanation in various areas of the urban environment will differ from location to location. The specific conditions that lead a resident of San Bernardino or Riverside to feel crowded may or may not apply to a city dweller in New York City. Physical and social conditions vary from one area to the next. However, it is clear that many situational factors may act to restrict the individual's freedom and control over the external environment. In this sense, the application of the present research must be limited to its general theoretical value, rather than as a specific prescription for urban planners.

A final note about the limitations of field research, in general, and the present study in particular should be included. Although this research has provided some interesting results, the ability to specify causal orderings of criterion and predictor variables is severely limited. In the case of many of the psychological variables, it is unclear whether these factors caused the perceptions of crowding, were a subsequent by-product, or were simply concurrent with the measure. Nevertheless, the value of this approach for consolidating a number of theoretical approaches and laboratory studies is apparent. It has contributed to an understanding of the phenomenon of crowding as it is manifested in an actual urban environment.

References

Altman, I. *The environment and social behavior*. Monterey, California: Brooks/Cole, 1975.

Brehm, J. *A theory of psychological reactance*. New York: Academic Press, 1966.

Buttimer, A. Social space and the planning of residential areas. *Environment and Behavior*, 1972, *4*, 279–318.

Carr, S. The city of the mind. In W. R. Ewald, Jr. (Ed.), *Environment for man: The next fifty years*. Bloomington: Indiana University Press, 1967.

Davis, K. The urbanization of the human population. *Scientific American*, 1965, *213*, 40–53.

Desor, J. A. Towards a psychological theory of crowding. *Journal of Personality and Social Psychology*, 1972, *21*, 79–83.

Esser, A. H. Experiences of crowding: Illustration of a paradigm for man–environment relations. *Representative Research in Social Psychology*, 1973, *4*, 207–218.

Freedman, J. L. *Crowding and behavior*. San Francisco: W. H. Freeman, 1975.

Freedman, J. L., Levy, A. S., Buchanan, R. W., & Price, J. Crowding and human aggression. *Journal of Experimental Social Psychology*, 1972, *8*, 528–548.

Glass, D. C., & Singer, J. E. *Urban stress*. New York: Academic Press, 1972.

Gochman, I. *Factors affecting the perception of crowding*. Unpublished doctoral dissertation, University of Washington, Seattle, 1976.

Hall, E. T. *The hidden dimension*. New York: Doubleday, 1966.

Kerlinger, F., & Pedhazur, E. *Multiple regression in behavioral research*. New York: Holt, Rinehart & Winston, 1973.

Kutner, D. H., Jr. Overcrowding: Human responses to density and visual exposure. *Human Relations*, 1973, *26*, 31–50.

Lee, D. H. K. The role of attitude in response to environmental stress. *Journal of Social Issues*, 1966, *12*, 83–91.

Milgram, S. The experience of living in cities. *Science*, 1970, *167*, 1461–1468.

Mitchell, R. Some social implications of high density housing. *American Sociological Review*, 1971, *36*, 18–29.

Proshansky, H. M., Ittelson, W. H., & Rivlin, L. G. Freedom of choice and behavior in a physical setting. In H. M. Proshansky, W. H. Ittelson, & L. G. Rivlin (Eds.), *Environmental psychology*. New York: Holt, Rinehart & Winston, 1970.

Rapoport, A. Toward a redefinition of density. *Environment and Behavior*, 1975, *7*, 133–158.

Saarinen, T. F. Perception of environment. Resource Paper #5. Washington, D.C.: Commission on College Geography, Association of American Geographers, 1969.

Saegert, S. Crowding: Cognitive overload and behavioral constraint. In W. Preiser (Ed.), *Proceedings of the environmental design research association* (Vol. 2). Stroudsburg, Pennsylvania: Dowden, Hutchinson, & Ross, 1973.

Schachter, S., & Singer, J. Cognitive, social, and physiological determinants of emotional states. *Psychological Review*, 1962, *69*, 379–399.

Schmidt, D. E., Goldman, R. D., & Feimer, N. R. Physical and psychological factors associated with perceptions of crowding: An analysis of subcultural differences. *Journal of Applied Psychology*, 1976, *61*, 279–289.

Schmidt, D. E., Goldman, R. D., & Feimer, N. R. Predicting perceptions of crowding at the residence, neighborhood, and city levels by physical and psychological measures. *Environment and Behavior*, 1979, *11*, 105–130.

Schneirla, T. C. The concept of development in comparative psychology. In J. Eliot (Ed.), *Human development and cognitive processes*. New York: Holt, Rinehart & Winston, 1971.

Sherrod, D. R. Crowding, perceived control, and behavioral aftereffects. *Journal of Applied Social Psychology*, 1974, *4*, 171–186.

Sonnenfeld, J. Variable values in space and landscape: An inquiry into the nature of environmental necessity. *Journal of Social Issues*, 1966, *12*, 71–82.

Stokols, D. On the distinction between density and crowding. *Psychological Review*, 1972, *79*, 275–277. (a)

Stokols, D. A social psychological model of human crowding phenomena. *Journal of the American Institute of Planners*, 1972, *38*, 72–83. (b)

Stokols, D. The experience of crowding in primary and secondary environments. *Environment and Behavior*, 1976, *8*, 49–86.

Stokols, D., Rall, M., Pinner, B., & Schopler, J. Physical, social and personal determinants of the perception of crowding. *Environment and Behavior*, 1973, *5*, 87–117.

Valins, S., & Baum, A. Residential group size, social interaction, and crowding. *Environment and Behavior*, 1974, *5*, 421–439.

Wicker, A. Undermanning theory and research: Implications for the study of psychological and behavioral effects of excess population. *Representative Research in Social Psychology*, 1973, *4*, 185–206.

Wohlwill, J. F. The physical environment: A problem for a psychology of stimulation. *Journal of Social Issues*, 1966, *12*, 29–38.

Worchel, S., & Teddlie, C. The experience of crowding: A two-factor theory. *Journal of Personality and Social Psychology*, 1976, *34*, 30–40.

4

Residential Density, Social Overload, and Social Withdrawal*

Dennis P. McCarthy and Susan Saegert

Introduction

A number of recent studies of the effects of high densities on people have suggested that it is not density *per se* that results in perceptions of crowding and negative social and affective responses. Rather these responses are related to the experience of excessive social encounters and exposure to more social information than a person can handle cognitively (e.g., Baum & Koman, 1976; Langer & Saegert, 1977; Saegert, MacKintosh, & West, 1975). These experiences are characterized by the term *social overload*.

The approach described emphasizes the importance of the particular form of the built environment, the activities and orientations of the people, and the social norms and bonds that exist. When these factors come together in such a way that individuals involved are subject to an environment that is difficult to structure and that exposes them to unpredictable and uncontrollable social interactions and interruptions of behavior, it is predicted that people will respond in an affectively nega-

*An earlier version of this paper was presented at the Eastern Psychological Association Conference, April, 1976. This paper was previously published in *Human Ecology: An Interdisciplinary Journal*, Vol. 6, No. 3, pp. 253–272. Copyright 1978 by Plenum Publishing Corporation.

Dennis P. McCarthy and Susan Saegert • Environmental Psychology Program, Graduate Center, City University of New York, New York, New York 10036.

tive and socially alienated way. High densities contribute to social and cognitive overload by increasing the number of other people with whom an individual may have to deal and by putting those people in close enough proximity that some experience of them is difficult for the individual to avoid. As experiences of each other's behavior become increasingly unpredictable, uncontrollable, and possibly interfering, the effects of high densities tend to become more detrimental. Further, the more time spent in the environment and the more diverse the activities engaged in, the greater the overload (Saegert, 1978).

The study to be described was designed to examine the effects of high density in the immediate residential environment on residents' experiences of social overload and the consequences of these experiences for their social relationships and attitudes. The study site, a low-income housing project, provides a location in which residents are maximally likely to experience overload: it is a large development, housing over 2,000 families from which the tenants, like most low-income people, have little opportunity to escape; an environment in which an atmosphere of uncertainty, fear, mistrust, and what Rainwater (1966) has described as "touchiness" in neighborly relations prevails.

The psychological importance for low-income people of the area immediately adjacent to their homes has been pointed out in many studies (e.g., Brower, 1975; Brower & Williamson, 1974; Fried & Gleicher, 1970; Gans, 1962; Rainwater, 1966, 1970; Schorr, 1966). These studies suggest that lower-income residents attempt to extend their sense of security and belonging beyond the immediate dwelling unit into the outer residential environment. The process seems to be one of identifying a larger residential group with whom one senses a certain amount of security and trust, and thereby a sense of stability and predictability in the area immediately beyond one's own four walls. Over time, feelings of commitment, responsibility, and a sense of belonging to the area and to those residents who share it seem to develop, if the sense of security and predictability can be maintained. These studies indicate that while working-class people in such traditional areas as the West End of Boston are able to develop these attachments (Fried & Gleicher, 1970; Gans, 1962), very low-income public housing residents, for the most part, are not (Rainwater, 1966).

Newman (1973) has characterized this problem as lack of defensible space. However, it appears to us that territorial defense is just one aspect of the more extensive problem of dealing effectively with a highly unpredictable and untrustworthy sociophysical environment. We see the number of individuals in the immediate environment as a crucial factor affecting the achievement of a sense of control in the setting and a more general sense of one's ability to manage life in the environment successfully.

In the setting of a low-income public housing project we would expect that high density would have very strong negative effects. The tenants would tend to experience overload because most of their daily living activities occur in this residential environment and thus a great amount of time is spent therein.

Public housing tenants are selected from long waiting lists and are assigned randomly to apartments as they become available; thus it is rare for any social bonds to exist prior to moving into the project. It would be expected from other research (Rainwater, 1966, 1970) that the tenor of life for most residents would involve anxiety about meeting basic needs, great concern about physical and social security, difficulties in raising children in the way they desire, and mistrust and fear of their neighbors.

The particular project selected for this study allows us to look at the differential effects of residential density within the same setting and for a homogeneous population. The project contains both high- and low-rise buildings, thus providing comparisons between people who share their immediate residential environment with 12 other families (in the low buildings) and those sharing them with 110 other families (in the high-rise towers). Our predictions were that high-rise residents would experience more social overload, which would result in more perceived crowding, feelings of less control, safety, and privacy in the residential environment, as well as less positive and more problematic social relationships with other tenants, and more feelings of withdrawal, alienation, and dissatisfaction with the residential environment generally.

The Study

Structured interviews were conducted during the summer of 1975 with tenants of a low-income public housing project in the northeast Bronx. The Bronx, with a population of 1.5 million, is the northernmost borough of New York City and is located between Manhattan and suburban Westchester County. The northern and eastern portions of the Bronx are suburban in nature while the southern and middle portions appear to be extensions of Manhattan. The borough's total land area is 41.5 square miles.

The area immediately surrounding the housing project contains a good deal of greenery. It consist of predominantly working-class and lower-middle-class single-, two-, and three-family homes. Slightly fewer than 2,000 families live in the project's 56 3-story walk-up and 12 14-story high-rise buildings. Of the project's 6,980 residents, 62% are black, 19% are white (most of whom are elderly, over 65 years of age), and 18% are Puerto Rican. The tenant population is typical of other low-income housing projects in New York City, and other than its large overall size and mix of building types, the project's physical design is representative of other low-income housing projects in the city. Construction of the development was completed in 1953.

Tenants were interviewed in six 3-story buildings and two 14-story buildings located within the same area of the project. The buildings in the study area were all equal distances from neighborhood stores, the management office, the housing police station, and the nearest neighborhood police precinct.

Most of the 14-story buildings, or high-rise towers, house 110 families, eight apartments on a floor, with the exception of the first floor where there are only six apartments (the high-rise building for the elderly contains somewhat more apartments but was excluded from our sample). The apartments on each floor open onto long double-loaded corridors and are reached by an elevator located in the middle of the corridor. The 3-story buildings are walk-ups housing 12 families per entrance in clusters of four apartments off each stairway landing.

Both the high- and low-rise building types have lobby areas on the ground floor where tenants' mailboxes are located. Also, immediately adjacent to each building's entrance are benches and children's play equipment designed for use by the residents. Apart from the fact that the low-rise apartments have three bedrooms, while the high-rise apartments have two, apartments within the two building types are very similar.

All the project residents are drawn from the same pool of applicants to the New York City Housing Authority, and except for the placement of slightly larger families in the 3-story apartments, which contain an extra bedroom, tenants are randomly assigned to buildings within the project. If family size increases, families living in the 14-story buildings may be transferred to a 3-story building when necessary. Self-selection of building type preference is not possible for tenants.

A total of 60 interviews were conducted, 30 in each building type. Interviews were conducted equally across the buildings chosen within each building type: that is, 15 tenants were interviewed in each of the two high-rise buildings and from 4 to 6 tenants in each of the six low-rise buildings chosen for study.

Three interviewers, one male and two female, in teams of two, conducted all of the interviews. Interviewers knocked at doors randomly and interviewed the female head of households. When necessary, interviews were conducted in Spanish. The actual interview, which took an average of 45 minutes, contained short-answer questions or questions eliciting responses along 6-point scales. Groups of scalar and short-answer inquiries were interspersed with open-ended questions, which as yet have not been completely analyzed. The interview attempted to elucidate the tenants' daily experiences in various interior spaces of their buildings as well as their social relations with others living on their floor, in their building, and in the project, and with close friends and relatives living outside the project. The interview also included measures of tenants' overall orientation to the project and to life in general.

Findings

Demographic comparisons (see Table 1) of the high- and low-rise apartment tenants interviewed revealed that the two groups did not differ significantly in age, education, car ownership, length of residence in the project, length of residence in present apartment, number of parents in the household, or number of children under age 12. The only significant demographic difference between the groups was in terms of family size: mean family size in the three-story buildings was 5.2 persons, approximately one member larger than families in the high-rise buildings. This difference was expected due to the difference in number of bedrooms.

Overall, comparison of living experiences of high- and low-rise tenants revealed that residents of the 14-story buildings were (1) more likely to report experiences of social overload and crowding; (2) more likely to feel a weaker sense of control, privacy, and safety in various interior

Table 1. Demographic Comparisons of High- and Low-Rise Apartment Tenants

	High-rise $\bar{X}$	Low-rise $\bar{X}$	Significance of t
Age	37.63	41.30	n.s.
Education (years)	11.76	11.20	n.s.
Cars per household	.53	.53	n.s.
Length of residence in project (years)	10.43	13.80	n.s.
Length of residence in present apartment (years)	9.93	7.83	n.s.
Number of parents per household	1.53	1.63	n.s.
Number of children under 12 per household	1.03	1.53	n.s.
Family size	3.57	5.27	$p < .001$

spaces of their building; (3) more likely to experience greater difficulty in their social relations; and (4) more alienated, less satisfied, less involved, and more detached from their own building and the project in general.

Social Overload and Tenant Anonymity.   It was hypothesized that tenants in the high-rise buildings would come into contact with large numbers of others in the public areas (hallways, elevator, and lobby) of their buildings. As these contacts exceeded the residents' interaction capacity or ability to process relevant incoming social stimuli, tenants would experience social overload. This experience would be manifested by tenants' perceptions of crowding in the building and reports of trouble in the building as relations between tenants became unmanageable. Contact with large numbers of others in those areas beyond the apartment was also expected to lead to a sense of anonymity among residents and to cause difficulty in tenants' discrimination of fellow residents from intruders.

These predictions were supported (see Table 2). Tenants in the high-rise apartments reported seeing people in the lobby and elevator more often, felt more crowded in their buildings, and felt that there was more trouble in their buildings than tenants in the three-story buildings. Measures of building anonymity also proved significant: high-rise apartment residents saw people they did not recognize in the lobby and elevator more often and knew a smaller portion of fellow residents to say "hello" to in the building than did low-rise apartment tenants.

Building Control, Safety, and Privacy.   High-rise apartment tenants' contact with large numbers of others in their buildings and the general anonymity among residents in these buildings might explain the

Table 2.   Social Overload and Tenant Anonymity

	High-rise $\bar{X}$	Low-rise $\bar{X}$	Significance of t
Social overload			
How often see people in lobby, elevator, or in stairs[a]	1.50	2.90	$p < .001$
Crowded in building[b]	2.43	4.77	$p < .001$
Tenant anonymity			
How often see people don't recognize in lobby, elevator, or in stairs[a]	3.03	4.67	$p < .001$
Proportion of tenants know to say "hello" to in building[c]	3.57	1.63	$p < .001$

[a]Six-point scale: 1 = very often; 6 = never.
[b]Six-point scale: 1 = very crowded; 6 = not crowded at all.
[c]Six-point scale: 1 = all; 6 = none.

Table 3. Control, Safety, and Privacy in Building Areas

	High-rise $\bar{X}$	Low-rise $\bar{X}$	Significance of t
Control[a]			
Elevator or stairs	5.10	2.93	$p < .001$
Lobby	4.97	3.00	$p < .001$
Safety[b] (in these areas at night)			
Hallway on floor	3.37	2.40	$p < .019$
Elevator or stairs	4.33	2.70	$p < .001$
Lobby	4.48	2.73	$p < .001$
Outdoor bench area	4.52	3.10	$p < .003$
Privacy[c]			
Hallway on floor[d]	4.10	3.27	$p < .05$
Elevator or stairs	5.40	3.57	$p < .001$
Lobby	5.63	3.53	$p < .001$

[a] 1 = a lot of control; 6 = no control at all.
[b] 1 = very safe; 6 = very unsafe.
[c] 1 = very private; 6 = very public.
[d] When family size was controlled this measure was no longer significant.

lower sense of control, safety, and privacy they reported feeling in building spaces beyond their apartments (Table 3). Residents of the 14-story buildings felt significantly less control over the elevator and lobby in their buildings than did low-rise tenants; they reported feeling significantly less safe in the hallway on their floors, in the elevators, in the lobbies, and on the benches in front of their buildings during the night; and they felt that the hallways, elevators, and lobbies lacked privacy significantly more than did low-rise tenants in these areas.

It is interesting to note that there were no significant differences between the high- and low-rise apartment tenants in their feelings of control, safety, and privacy in their apartments. Also interesting, although less marked in the low-rise buildings, is the generally uniform trend of decreasing feelings of control, safety, and privacy in both building types as one moves further away from the immediate apartment.

Informal Social Relations. Informal social relations with people inside and outside their buildings also differed significantly between high- and low-rise tenants. Both groups reported similarly low numbers of close friends and relatives living in their buildings, in the project, and outside of the project. High-rise tenants, however, engaged in less active socialization with those close friends and relatives living in their buildings (see Table 4). Residents of the 14-story buildings also reported greater doubt that a tenant in their building would come to the aid of another tenant being attacked in the building, or that tenants in their

Table 4.   Informal Social Relations

	High-rise $\bar{X}$	Low-rise $\bar{X}$	Significance of t
Social activity with others living outside the building[a]			
How often see, visit, go out with close friends living outside the project	3.48	2.19	$p < .003$
How often see, visit, go out with close relatives living outside the project	3.32	2.42	$p < .033$
How often visit with tenants living in the project	5.33	4.67	$p < .045$
Mutual aid[b]			
Proportion of tenants in building could count on for small favor	4.37	3.20	$p < .004$
Proportion of tenants in project could count on for small favor	5.30	4.57	$p < .008$
Proportion of tenants in building could count on in an emergency	4.10	3.23	$p < .034$
Territoriality[c]			
How likely tenant would come to aid of tenant being attacked in the building	3.62	2.45	$p < .015$
How likely tenant would interfere in act of vandalism to the building	3.38	1.97	$p < .006$
Use of outdoor bench area			
How often do you have casual conversations on the bench[a]	4.77	3.33	$p < .002$
How often in warmer weather do you sit on the bench[d]	5.40	3.90	$p < .013$

[a] 1 = very often; 6 = never.
[b] 1 = all; 6 = none.
[c] 1 = very likely; 6 = very unlikely.
[d] 1 = 4 or more times/week; 6 = never.

buildings would interfere in an act of vandalism to the physical plant of the building. Similarly, the high-rise apartment tenants reported significantly fewer tenants they could count on for a small favor in their buildings and in the project, and significantly fewer tenants in their buildings they could count on in an emergency.

Preliminary correlational analysis revealed a significant correlation of .52($N = 60$; $p < .001$) between sitting on the outdoor benches and casual conversation on the benches. Perhaps the relative absence of mutual social alliances among high-rise apartment tenants is related to the difference in the use low- and high-rise apartment tenants make of

Table 5. General Residential Orientation

	High-rise	Low-rise	Significance of t
Satisfaction with project and building Cantril Ladder			
(1) Project worst possible home; (10) project best possible home	$\bar{X} = 3.90$	$\bar{X} = 5.17$	$t = 2.00$ $p < .05$
Building preference			
(1) Prefer own building or undecided	43.3%	93.3%	$\chi^2 = 17.32$
(2) Prefer other building type	56.7%	6.7%	$p < .001$
Residential powerlessness "I don't feel there is much I can do to affect decisions made by the management":			
(1) strongly agree; (6) strongly disagree	$\bar{X} = 2.03$	$\bar{X} = 3.59$	$t = 3.34$ $p < .001$
Residential attachment			
"Belong just to my apartment"	40.0%	17.2%	$\chi^2 = 5.62$
"Belong to the whole project"	3.3%	24.1%	$p < .02$

their buildings' outdoor bench area. High-rise tenants reported making significantly less use of their immediate outdoor benches for casual contact with neighbors.

Residential Satisfaction, Detachment, and Powerlessness. Tenants' general orientation to their residential environment also differed greatly: those in high-rise apartments were significantly more dissatisfied with the project and their buildings; they reported a greater sense of powerlessness in affecting decisions made by the management, and a greater proportion felt detached from the project as a whole (See Table 5).

An interesting finding that may be related to the sense of powerlessness in affecting management decisions concerns the type of voluntary club and organizational membership reported (see Table 6). While there was a trend toward lower membership in voluntary clubs and organizations by tenants in the 14-story buildings, it is interesting to note that those clubs and organizations to which high-rise residents did belong were predominantly nonpolitical in nature. Sixty percent of tenants in low-rise apartments who belonged to voluntary organizations reported membership in such organizations as the Democratic Club, community health advisory boards, the Tenants Association, and the N.A.A.C.P., while only 25% of high-rise apartment club members belonged to such groups. It is plausible that high-rise apartment tenants' sense of powerlessness in their residential environment may come to

Table 6.   Residential Powerlessness and Voluntary Organization Membership

	High-rise	Low-rise	Significance
Residential powerlessness "I don't feel there is much I can do to affect decisions made by the management": (1) strongly agree; (6) strongly disagree	$\bar{X} = 2.03$	$\bar{X} = 3.59$	$t = 3.34$ $p < .001$
Voluntary organization membership Member of voluntary organization	27.6%	51.7%	$\chi^2 = 3.44$
Not a member of any voluntary organization	72.4%	48.3%	$p < .10$
Percentage of members who belong to a voluntary organization of a social-political efficacy nature	25.0%	60.0%	

affect their general sense of social and political efficacy in larger community and world matters.

Absence of Significant Apartment and Floor Differences. Another very interesting trend in the results is the relative absence of significant differences between the two groups on questions dealing with their apartments and the hallways on their floors (see Table 7). Tenants' residential experiences and perceptions of their floors in the 14-story buildings were in many ways similar to those of tenants in low-rise buildings, with the exception of sense of privacy and sense of safety in the hallway on their floors, which in comparison to other significant building effects were weak. The groups do not differ significantly concerning how often they see people in the hallway, how crowded they feel the hallway to be, how often they see people they do not recognize in the hallway on the floor, a sense of control over the hall, tenants they know to say "hello" to on the floor, whether a tenant on the floor would come to the aid of another tenant being attacked or would interfere if the building were being vandalized, and the number of tenants on the floor they could count on for a small favor or in an emergency.

Questions dealing with sense of control, privacy, and safety in the apartment also did not differ significantly between high- and low-rise apartment tenants.

General Alienation and Locus of Control. In order to investigate tenants' sense of general alienation in life, the Middleton Alienation scale and the short version of the Rotter Internal-External Locus of Control scale were included in the interview. Both high- and low-rise apartment residents' responses to these scales showed a similarly moderate level of life alienation and low internal sense of control.

Table 7. Significance of Differences between High- and Low-Rise Means for Apartment, Floor, and Building Questions (*t* Tests)

	Apartment questions	Floor questions	Building questions
See people	—	n.s.	$p < .001$
Crowded	—	n.s.	$p < .001$
See people don't recognize	—	n.s.	$p < .001$
Tenants know to say "hello"	—	n.s.	$p < .001$
Control	n.s.	n.s.	$p < .001$
Safety	n.s.	$p < .05$	$p < .001$
Privacy	n.s.	$p < .02^a$	$p < .001$
Tenant attacked	—	n.s.	$p < .02$
Vandalism to building	—	n.s.	$p < .006$
Tenants favor	—	n.s.	$p < .004$
Tenants emergency	—	n.s.	$p < .03$

[a]When family size was controlled this measure was no longer significant.

Family Composition.   To ensure that the primary factor mediating differences between high- and low-rise sample responses was building type and not variations in family composition, analyses were performed on those dependent measures that correlated significantly with family size and number of parents per household.

In testing the effects of family size, as no very large families live in the high-rise buildings and no very small families live in the low-rise buildings, tenants were first placed in subgroups based on family size and building type (high-rise apartment families of 2 to 3 members, high-rise apartment families of 4 to 5 members, low-rise apartment families of 4 to 5 members, and low-rise apartment families of 6 to 12 members). A series of 11 one-way analyses of variance, family size by building type, were then run for those dependent measures correlating significantly with family size. All of these dependent measures (see people in the lobby and elevator or stairs; control, privacy, and safety of elevator or stairs; control, privacy, and safety of the lobby; proportion of tenants know to say "hello" to in the building; vandalism to the building; preference of building type) continued to differ at the previous t-test significance levels with the exception of privacy of the hallway.

Following the analyses of variance, three subgroup comparisons were then performed for each of the 10 remaining significant measures. As expected, comparisons of high-rise apartment families of 2 to 3 members with high-rise apartment families of 4 to 5 members, and comparisons of low-rise apartment families of 4 to 5 members with low-rise apartment families of 6 to 12 members did not yield significant differences, while comparisons between high-rise apartment families of 4 to 5 members with low-rise apartment families of 4 to 5 members did yield significant mean differences for each of the 10 variables.

Analyses of covariance were then performed on the variables (proportion of tenants in project could count on for a small favor; proportion of tenants in the building could count on in an emergency) that were found to correlate significantly with number of parents per household. When the two analyses of covariance (high-rise–low-rise with effect of number of parents per household covaried out) were run, both of the dependent variables yielded differences between high- and low-rise apartments that were slightly more significant than the previous t tests had revealed, $F(1, 59) = 7.97$, $p = .006$; $F(1, 59) = 4.95$, $p = .028$, respectively.

Summary of Results

In sum, findings revealed that tenants interviewed in both building types were highly similar people. They did not differ demographically, in numbers of close friends and relatives, or in their general life alienation or sense of internal–external locus of control. It should be recalled that with the exception of sense of hallway safety, tenant responses to questions about their floors and their apartments did not differ significantly.

Significant differences were found in perceptions of and experiences in the building beyond the apartment and floor. High-rise apartment residents reported seeing people more often in the lobby and elevator and feeling more crowded in the building (our measures of social overload). They reported greater building anonymity, felt less control, privacy, and safety in interior public spaces, and had greater difficulty establishing mutually supportive neighbor relations beyond their own floors. High-rise tenants were less socially active, less satisfied with their buildings and the project, and felt more detached from the project as a whole. They also reported a greater sense of powerlessness in affecting management decisions and belonged to fewer voluntary associations in general than did low-rise apartment tenants.

Perceptions of Crowding

These results indicate vast differences in the everyday experience of tenants in the two building types. However, we were interested in pursuing further the question of to what extent these differences are related to the experience of crowding. Correlational analyses relating crowding to the various other measures therefore were undertaken.

Since it is unlikely that perceptions of crowding caused building type or floor height of residence, the two strong correlations between these variables and perceived crowding in the building suggest that these architectural factors lead residents to feel more crowded. The correlation of building type (treated as a dichotomous variable) with rated crowding was .55 ($N = 60$; $p < .001$) with high-rise apartment residents feeling more crowded.[1] Those living on higher floors also felt more crowded in the building; the correlation between these two variables was .59 ($N = 60$; $p < .001$). In interpreting this correlation it is important to remember that residents could not select either building type or floor height of residence.

Perceptions of crowding correlated significantly with most of our major dependent measures concerning (1) social overload and tenant anonymity, (2) building control, safety, and privacy, (3) informal social relations, and (4) residential satisfaction, detachment, and powerlessness. Residents who rated the building as more crowded also felt that they saw more people in the lobby and stairs, indicating that perceived

---

[1]The use of Pearson r in this manner, because of the computer program used, provides the same results as would the biserial r (Nunnally, 1967).

crowding is related to social overload. Further, perceptions of crowding were related to indications that residents could not deal effectively with the presence of so many people. Those who reported going out of their way to avoid other tenants felt more crowded. Perceived crowding correlated significantly with seeing people whom one did not recognize and also with perceptions of trouble among tenants.

In addition, this analysis showed a relationship between perceptions of crowding on the floor and in the building. Table 8 presents these correlations.

Multiple and partial correlations were performed to further investigate the relationships among the physical variable of building type, the experiences of seeing people, and perceptions of crowding and of trouble between tenants. The correlation between (a) building type (used as a dichotomous variable) and (b) perceived crowding is .55, explaining 30% of the variance. When a multiple partial correlation is performed in which both experiences of (c) seeing people (in the lobby and elevator or stairs) and of (d) seeing people who are not recognized are controlled for, the relationship between perceived crowding and building type is greatly reduced $[r(a)(b) \cdot (c)(d) = .30]$, explaining now only 9% of the variance. Thus it appears that experiences of overload do mediate the greater perception of crowding in high-rise buildings.

Further, another set of analyses suggest that perceptions of crowding mediate the relationship between building type and perceived trouble among tenants. The correlation between (b) perceived crowding and (y) perceived trouble among tenants is .56, explaining 31% of the variance. When (a) building type is partialed out of this relationship the variance explained is somewhat reduced $[r(b)(y) \cdot (a) = .42$; variance explained $= .17]$. Similarly, controlling for perceptions of trouble reduces the relationship between perceived crowding and building type some-

Table 8. Correlations of Social Overload
Measures with Perceived Crowding

	r	N	p
Frequency of seeing people in lobby and elevator/stairs	.48	60	<.001
Frequency of seeing people do not recognize	.49	60	<.001
Try to avoid others	.24	60	<.04
Feel that there is trouble among tenants	.56	51	<.001
Crowded floor	.26	60	<.02

Table 9.   Correlations of Ratings of Control,
Safety, and Privacy and Perceived
Crowding

	r	N	p
Control of stairs/elevator	−.50	60	<.001
Control of lobby	−.53	60	<.001
Safety in hall[a]	−.43	60	<.001
Safety on stairs/elevator[a]	−.50	57	<.001
Safety in lobby[a]	−.55	58	<.001
Safety in bench area[a]	−.50	59	<.001
Privacy of hall	−.28	60	<.01
Privacy of stairs/elevator	−.45	60	<.001
Privacy of lobby	−.56	60	<.001
Privacy of bench area	−.34	60	<.004

[a] At night.

what $[r(b)(a) \cdot (y) = .40$; variance explained $= .16]$. However, the largest reduction of relationship occurs when perceived crowding is partialed out of the correlation of building type and perceived trouble $[r(a)(y) \cdot (b) = .20$; variance explained $= .04]$.

These partial and multiple partial correlations begin to suggest some of the interrelationships among variables that are of interest to us, and along with techniques of path analysis they will be employed more extensively to data now being gathered by the authors. The sample will be expanded, as will the number of sites studied. This larger sample will permit data reduction through factor analysis prior to the attempt to construct a possible model of causal relations.

Significant correlations also were obtained between perceived crowding in the building and many measures of feelings of control, safety, and privacy (shown in Table 9). Rated control of the elevator or stairs and of the lobby were strongly related to perceived crowding, as were feelings of safety in and perceptions as public of, the bench area, lobby, stairs or elevator, and hall.

The network of feelings and perceptions in which perceptions of crowding are embedded extends to social relationships in the building, the project as a whole, and even to relationships outside (see Table 10). Residents who feel that the building is more crowded also perceive the other residents as less friendly, less likely to do small favors, less likely to stop vandalism, and less likely to intervene if another tenant were attacked. They know fewer people in the building to whom they say "hello" and also engage in less casual conversation in the bench area outside the building. Some of these same orientations of those who feel

Table 10. Correlations of Measures of Social Relationships and Perceived Crowding

	r	N	p
Think tenants in building are friendly	−.25	56	<.03
How many people know in building to say "hello"	−.41	60	<.001
How many people know in project to say "hello"	−.23	60	<.04
Frequency of casual conversation in bench area	−.26	60	<.02
Probabilty another tenant would intervene in attack in building	−.33	54	<.007
Probability another tenant would stop vandalism in building	−.30	58	<.01
Number of tenants in building could count on for small favor	−.24	60	<.03
Number of tenants in project could count on for small favor	−.22	60	<.04
Frequency visit friends living outside project	−.29	51	<.02

more crowded extend to the project generally. Ratings of crowding are correlated with knowing fewer people to say "hello" to in the project as a whole and feeling that there are few people in the entire project who would do a small favor for them. Further, those residents who rate their buildings as more crowded are less likely to visit friends living *outside* of the project.

Perceptions of crowding also relate to the attitudes residents have toward the residential environment (Table 11). Our measures of residential powerlessness and estrangement both correlate with crowding ratings, as do tenants' ratings of residential satisfaction and their desire to stay or move. Our speculation that the effects of social overload ir-

Table 11. Correlation of Measures of Alienation, Detachment, and Residential Satisfaction with Perceived Crowding

	r	N	p
Residential powerlessness	.30	59	<.01
Residential estrangement	.40	60	<.001
Number of clubs/organizations	−.24	60	<.03
Time spent inside apartment	.29	60	<.01
Time alone inside apartment	.30	60	<.01
Residential satisfaction (Cantril Ladder)	−.36	59	<.003
Desire to move	.31	60	<.008

radiate into other areas of the residents' lives is supported by the correlation between rated crowding and membership in clubs or organizations. Further, it appears that those people who feel more crowded in the building withdraw by spending more time inside their apartments and that more of this time is spent alone.

As in other studies on the effects of density (e.g., Mitchell, 1971), those residents who feel more crowded also report more difficulty keeping track of their children. Although the sample of residents with children under age 12 is small, we found a correlation of .48 ($N = 29$; $p < .004$) between these two measures.

The seriousness of the impact of social overload on residents is further suggested by the correlation of .24, significant at the .03 level, between length of residence and perceived crowding. Thus not only do residents *not* adapt to their conditions over time, but the experience of crowding tends to become slightly stronger. It is important to note that none of the other demographic variables was correlated with perceptions of crowding, once again indicating that *environmental*, not *personal*, differences are the major determinants of perceived crowding in this study.

These findings indicate that crowding is related to a whole complex of perceptions of social overload, lack of control, safety and privacy, social isolation and withdrawal, and negative evaluations of the residential environment. Certainly it could be argued that all of these measures simply relate to antisocial personality traits. However, the extremely strong correlations between perceptions of crowding and two architectural features, building type and floor height, suggest that environmental factors are important in determining these negative orientations and feelings. Further, almost all of the measures that correlate significantly with perceived crowding also revealed significant mean differences as a function of building type. Therefore, it seems reasonable to conclude that the network of feelings, attitudes, and social behaviors related to perceived crowding arises for these residents from conditions of social overload, brought about by the unmanageably large number of unrelated residents brought together in the high-rise buildings, and that these effects are stronger for those who have more occasion to come in contact with the numerous other residents because of occupancy of apartments on upper floors.

Conclusions

Perhaps the most outstanding feature of the data gathered in this study is the strength and reliability of the findings. Of the many questions asked, practically all of those that we expected to be related to social overload showed significant differences between the two building types. The correlational analysis further supports the hypotheses. The results of this study are in keeping with the studies by Baum and Valins (1975) at the State University of New York at Stony Brook showing that social overload in a dormitory setting related to felt crowding as well as social withdrawal and more negative social attitudes. However, our study, since it focused on people whose lives are considerably more insecure and difficult than those of most college students, reveals more clearly the extremely severe and alienative effects that social overload can have in situations that are already fraught with unpredictability and lack of control. In fact, in light of other studies of public housing (cf. Rainwater. 1966), the really surprising information in our study concerns the extent to which residents in the low-rise buildings were able to build relationships of trust and mutual aid.

The evidence that residents in the high-rise buildings neither compensate by spending more time away from the project nor having relationships with people elsewhere nor adapt to the experience of overload in the course of time also emphasizes the seriousness of the effects of high residential density. In contrast to theories suggesting that people adapt to or cope with stresses over time, our data indicate that perceptions of crowding may become stronger the longer one lives in a crowded environment and that the social withdrawal engendered by social overload in the residential environment extends to other aspects of life. Our findings that high-rise apartment residents visit friends away from the project *less* frequently than do low-rise apartment residents and that they belong to fewer organizations, especially political organizations, suggests that the experience of overload in such a personally significant area as that immediately adjacent to the home can dispose individuals to avoid social involvement generally. Again, these data corroborate and extend the finding of Valins and Baum (1973) that crowded dormitory residents behaved in a less prosocial manner in a laboratory setting as well as in their dormitories.

This study also strongly supports the hypothesis that the negative effects of high densities are mediated by social overload: the correlations between perceived crowding and our measures of social overload were all in the mid-40s and 50s. Unlike the Baum and Valins work, it also shows the link between social overload and higher densities, at least when higher density involves greater numbers of others in the immediate residential environment. Together, these studies raise the question of the extent to which building design can alleviate the experience of social overload, even when densities are rather high. Extending the implications of the Baum and Valins studies and adopting some of the design recommendations, but not the theoretical underpinnings, of

Newman's analysis, we would expect that building designs that created manageable subgroups of residents would lead to less probability of social overload.

One of the intriguing issues raised for designers as well as social and cognitive psychologists concerns the determinants of a psychological unit and the functions of such units. Clearly, all our respondents viewed their buildings as a unit of some kind. For the low-rise apartment residents, this was a unit to which they belonged, in which they felt some sense of trust, social involvement, and responsibility, and which served as a base from which to extend their social commitments into the outer world. In contrast, high-rise apartment residents saw the building as a dangerous conglomerate of alien spaces and mainly threatening people. They made some differentiation between their own floors and the rest of the building, but their orientation toward the building did flow over into their feelings about the floor to the extent that they felt less safe in the hall than did low-rise apartment tenants. These tenants made it clear that their major identification was with their own apartments. They neither identified with nor took responsibility for the people and spaces of the building or the project as a whole. In fact, withdrawal from external social contacts characterized their responses to a wide range of questions.

The implications of these findings are that designers must take seriously the importance of creating environmental units with which residents can identify. Critics of the notion that high densities can lead to negative experiences and antisocial relationships have suggested that even in high-density environments people make their own manageable social units by relating to neighbors, friends, and relatives. For the population we studied, this clearly does not occur; such supportive groups in our study grew up only when the physical environment facilitated casual contact with and personal identification of others in the building. Any design solution expected to facilitate such relationships must induce cognitive categories and manageable numbers and types of interactions for the residents. The site of our study well exemplifies the way in which inaccurate or overly simplified psychological constructs can lead to useless design interventions, having been the site of interventions based on the idea of creating "defensible space" for residents. Neglect of the importance of cognitively manageable physical units, and lack of awareness of the network of social attitudes and relations in which territorial defense exists, led to the putting up of fences around various areas of the project. The overload perspective would indicate that such interventions are on too large a scale to be very helpful to residents. Informal comments from respondents suggest that residents may find the fences irrelevant and very often irritating.

From the point of view of some critics of high-rise buildings (e.g., Yancey, 1971), our findings might be taken as evidence for the isolating qualities of high-rise buildings and thought to be unrelated to social overload. However, the large differences in our overload measures between the high- and low-rise samples and the correlations between perceptions of crowding and our other dependent measures suggest that such a criticism is unwarranted and that the isolation these researchers have observed may in fact stem from the social overload occasioned by such buildings.

The major limitation of our data arises in any attempt to generalize our findings to middle- and upper-income groups. Baum and Valins's studies of college students suggest that some generalization is possible. However, middle- and upper-income adults would be expected to have more opportunities to choose residential environments with which they identify and to live with people with whom they already have social ties, ranging from their families to compatible neighbors. The possible role of perceived control, safety, and privacy in mediating the negative effects of social overload may not be critical for higher-income groups, whose perceptions of these factors may be less affected by the number of others encountered in the environment. Even if some of the experiences and problems associated with social overload did exist for these populations, we would expect that adaptation and coping would be more likely to succeed in ameliorating the situation. Clearly, further research in this direction is needed.

References

Baum, A.. Harpin, J., & Valins, S. The role of group phenomena in the experience of crowding. *Environment and Behavior*, 1975, *7*, 184–198.

Baum, A., & Koman, S. Differential response to anticipated crowding: Psychological effects of social and spatial density. *Journal of Personality and Social Psychology*, 1976, *34*, 526–536.

Brower, S. Recreational uses of space: An inner city case study. *Social Ecology*, 1975, *3*, 153–166.

Brower, S., & Williamson, P. Outdoor recreation as a function of the urban housing environment. *Environment and Behavior*, 1974, *6*, 295–345.

Fried, M., & Gleicher, P. Some sources of residential satisfaction in an urban slum. In H. Proshansky, W. Ittelson, & L. Rivlin (Eds.), *Environmental psychology*. New York: Holt, Rinehart & Winston, 1970, pp. 333–346.

Gans, H. *The urban villagers*. New York: Free Press, 1962.

Langer, E., & Saegert, S. Crowding and cognitive control. *Journal of Personality and Social Psychology*, 1977, *35*, 175–182.

Mitchell, R. E. Some social implications of high density housing. *American Sociological Review*, 1971, *36*, 18–29.

Newman, O. *Defensible space*. New York: Collier Books, 1973.

Nunnally, J. C. *Psychometric theory*. New York: McGraw-Hill, 1967.

Rainwater, L. Fear and the house-as-haven in the lower class. *Journal of the American Institute of Planners*, 1966, *32*, 23–31.

Rainwater, L. *Behind ghetto walls*. Chicago: Aldine, 1970.

Saegert, S. High density environments: Personal and social consequences. In A. Baum & Y. Epstein (Eds.), *Human responses to crowding*. Hillsdale, New Jersey: Erlbaum, 1978.

Saegert, S., MacKintosh, E., & West, S. Two studies of crowding in urban public spaces. *Environment and Behavior*, 1975, *7*, 159–184.

Schorr, A. *Slums and social insecurity*. Washington, D.C.: U.S. Government Printing Office, 1966.

Valins, S., & Baum, A. Residential group size, social interaction, and crowding. *Environment and Behavior*, 1973, *5*, 421–439.

Yancey, W. Architecture, interaction and social control. *Environment and Behavior*, 1971, *1*, 3–21.

5

Density, Perceived Choice, and Response to Controllable and Uncontrollable Outcomes*

Judith Rodin

With increasing density in almost every major urban center and the population continuing to grow dramatically, the study of crowding has begun to attract considerable attention in the social sciences. Interestingly, it was a series of animal studies (Calhoun, 1961, 1962) that first provided systematic evidence associating a number of social pathologies such as infanticide, aberrant sexuality, and high rates of mortality with increased density. In his experiments, Calhoun simply provided normal rats with sufficient food and water and allowed them to reproduce and overpopulate in a fixed area. The pathologies developed when high densities were reached. While there are correlational studies that have reported comparable social disorganization and pathology related to overcrowding in humans (Galle, Gove, & McPherson, 1972; Schmitt, 1957, 1966; Winsborough, 1965), Freedman (1975) has suggested that the statistical relationship may occur because both are effects of poverty and

*This paper was originally published in the *Journal of Experimental Social Psychology*, Vol. 12, pp. 564-578. Copyright 1976 by Academic Press, Inc., New York. New York. Reprinted by permission.

Judith Rodin • Department of Psychology, Yale University, New Haven, Connecticut 06520.

undereducation and concluded that high density does not necessarily have negative effects on people (Freedman, Heshka, & Levy, 1975).

Experimental paradigms to study the effects of overcrowding can generally be classified into two basic types. Short-term manipulations have varied numbers of persons and room size and measured perceptions of crowding, physiological arousal, task performance, aggression and/or competition, and interpersonal relations (see, for example, Aiello, Epstein, & Karlin, 1975a; Freedman, 1975; Hutt & Vaizey, 1966; Saegert, MacKintosh, & West, 1975). Some studies have shown that density is positively related to increased physiological arousal (Aiello *et al.*, 1975a) and aggression and discomfort (Hutt & Vaizey, 1966; Saegert *et al.*, 1975), while others have suggested that density simply intensifies the dominant feature of the situation, whether positive or negative (Freedman, 1975). In a second type of procedure, experimenters have varied the amount of usable space per person for longer periods of time and again reported conflicting results from study to study. Smith and Haythorn (1972) confined two- and three-man groups to a small or large space for 21 days and found that less hostility was expressed in the smaller room than in the larger. On the other hand, Aiello *et al.* (1975b) and Baron, Mandel, Adams, and Griffen (1975) tested the effects of tripling in dormitory rooms designed and intended for two people and found large increases in negativity of reactions when a third person was added.

In the studies described in the present report, a third type of experimental approach was used to test the effects of high density. Subjects were either selected on the basis of their residential density or randomly sampled and subsequently tested for the effects of residential density. The correlates of these natural variations in density were measured in the laboratory, where experimental manipulations were designed to test hypotheses regarding the intervening processes by which high-density living might affect behavior. Specifically, we hypothesized that one of the more serious consequences of high-density living may be a real or perceived inability to control the environment and to regulate one's social interactions. Proshansky, Ittelson, and Rivlin (1970) and Zlutnick and Altman (1972) have suggested that control over what goes on in defined areas of space and over interpersonal exchange allows an individual to perceive that he has or can achieve some freedom of choice. If this line of reasoning is correct, and assuming that high-density living may lessen an individual's opportunity to exercise control, one would expect to find important differences between residents of high- and low-density settings in their response to outcome control and freedom of choice.

Experiment 1

The first study examined the effects of chronic high-density living on attempts to control the selection of available outcomes. Experiments have demonstrated that subjects typically prefer or will work harder for positive (Brigham & Sherman, 1971) and negative (Pervin, 1963) reinforcement that they, rather than the experimenter, administer. This occurs even when the actual shock or reward is the same regardless of who administers it. If high density minimizes the opportunity to control the onset of important outcomes, the effects of this experience should be reflected in other control-relevant behaviors. On the one hand, it is possible that adaptation to chronic high density and uncontrollability leads to less choice behavior in any situation. On the other hand, it is possible that individuals living under these conditions try harder to control their reinforcement whenever they perceive the opportunity to do so. To determine which of these possibilities was true, we used a variant of an operant conditioning procedure developed to determine the extent to which the opportunity to choose between self- and experimenter-administered positive outcomes was reinforcing (Brigham & Sherman, 1971).

The Study

Subjects. Subjects were 32 black males between the ages of 6 and 9 years old, living in the same multiapartment low-income housing project. All lived in three-room, one-bathroom apartments that were identical in interior layout. To select these subjects, an interview was conducted about 3 months before the experiment by a black 20-year-old male who lived in the housing project. He asked a predetermined series of questions to assess the number of people living in each apartment and their length of time in residence. He also asked about the age, family relationship, and occupation of all residents. On the basis of this interview we were able to select as subjects male children who lived with two through eight other people. Each had to have lived with that number of persons for at least 2 years in order to qualify.[1]

[1]In federally subsidized low-income housing, rent is a function of the number of rooms and income and, in theory, the amount of space allotted is a function of the number of people in the family unit. In practice, however, this is rarely the case. The number of people in the extended family or nonrelations living in each unit is typically not reported. Similarly, information about children or other family members leaving the home is rarely given to the housing authority. Increase in family size, although sometimes reported, simply secures a place on a long waiting list for a larger apartment. Consequently, the family unit living in these three-room apartments contained from 3 to 10 people.

Whenever there is nonrandom assignment of subjects, sampling procedures become a critical issue. Although several stringent criteria were developed on which to base subject selection, they do not guarantee that the subjects actually tested were identical in all aspects except the density of their homes. In each condition a variety of family structures were represented, but they were not equally represented. For example, in larger groups it was generally true that there were more children. However, a subject's birth order was not systematically related to group size. The number of first-borns was about the same in each group, with one first-born in groups of size two, six, and eight; two in groups of size three, five, and seven; and three in the group of size four. The number of unrelated adults with whom the child lived increased with group size, but since the number of siblings also increased, the adult–child ratio was, on the average, fairly constant across conditions. There were no differences among conditions in the number of two-parent families living together. Some children in the smaller groups lived with a mother and sibling(s) but no father, while another lived with a sister and two unrelated adults. Similarly, some children in larger size "families" lived with unrelated or distantly related adults. Since subjects were contacted and recruited at their elementary school, their IQ test scores were available to us. While correlations between *family* size and IQ usually run $-.30$ (cf. Anastasi, 1958), there was no statistically significant correlation between *household* size and IQ, $r(27) = -.16$.

The interviews also revealed that the larger groups did not have significantly more disruption and instability (in terms of movement in and out of the family unit) than the smaller ones. There was some tendency for larger units to report earning more money than small families; however a median split comparing groups of sizes two, three, four, and five to groups of size six, seven, and eight, below and above the average per-person income, showed no significant difference, $\chi^2(1) = .831$.

However, there were differences in the number of screening interviews required to find subjects from different size household groups. More interviews were needed to find children at the extremes, as indicated in Table 1. There were three additional subjects, one in each of groups five, six, and seven, who were not included in the data analysis because postexperimental questioning indicated that they had not fully understood the experimental procedure. While this description of similarities and differences among subjects in the various conditions is not exhaustive, it is intended to provide greater detail on at least some of the relevant characteristics of the sample.

Apparatus.   The apparatus was a white box measuring $12 \times 24 \times 12$ inches with a face of opaque Plexiglas and a plywood strip near the

Table 1. Total Number of Families Interviewed by Condition

Number of people living with child	Total number interviewed	Number in sample
2	20	3
3	14	5
4	13	5
5	12	4
6	12	4
7	18	3
8	17	4

base. Mounted on the plywood were two response buttons, one at the center and one 3 inches to the right of center. The plastic screen could be illuminated by red or green bulbs mounted behind it. Two Gerbrands feeders, one for candy and the other for marbles, were attached to the box and delivered candy and marbles to the child into a clear plastic tray at the bottom of the base of the apparatus. Schedules were switched and responses recorded automatically.

Procedure. Children were recruited through their school where the actual testing took place. The experimenter was a black female who was unaware of a subject's residential density. At the beginning of the first session, she showed each child the apparatus and described how it worked. The experimenter showed him that when he pressed the button at the center of the panel, either candy or marbles would drop into the tray as a reward for pressing. She told him that the color of the light would be the signal to tell him whether his reward would be candy or marbles at any particular time. After each subject had earned two pieces of candy by pressing the button,[2] the experimenter described the procedure for the first test. Every time a new schedule was introduced, the child was informed whether candy or marbles would be earned for pressing the button. A red or green light served as the discriminative simulus for the nature of the reward.

Each trial was 6 min long, consisting of 12 alternations between a 30-sec red light period and a 30-sec green light period. Reinforcement was delivered on an FR-35 schedule (a fixed ratio of 35 responses were required to produce either a piece of candy or a marble) separately programmed for the responses on each side. Ten trials per day were given on each of 3 days spaced 2 to 3 days apart.

[2]A large assortment of candy, such as Tootsie Rolls, peanut chews, and wrapped chocolate pieces, was used.

Multiple Schedule Procedure. In all trials, *immediate candy* was earned for responses when the signal was one color and *marbles* for responses to the other light signal. For half the children in each group the red light signaled immediate candy and the green light signaled marbles; for half, the opposite color light signaled marbles; for half, the opposite color light–outcome relationships were conditioned.

Day 1. On trials 1–5 the earned marbles could not be traded for candy. After a short break, the new schedule was described and on trials 6–10 marbles could be traded for *experimenter-selected* candy. The color of light signal representing the two outcomes (immediate candy or marbles) was also switched.

Day 2. Trials 11–15 were identical to trials 1–5. The only change was the color light that signaled the two outcomes. After a short break and a description of the new schedule, trials 16–20 were begun. Marbles could be traded for *self-selected* candy. Again, the color of light signaling the two outcomes was switched.

Concurrent Schedule (Day 3). The new procedure allowed the child himself to switch from one component of the schedule to the other at any time. This was done by activating the second key. The switching key was introduced by demonstration; the experimenter simply pressed it several times, showing the child that pressing the key changed the stimulus condition from green to red or red to green. The child was then asked to press the key a number of times and told that he could do so as often as he liked during the session.

A press to the switching key produced 30 sec of a component, terminated by an automatic return to the alternate component. The signaled components alternated every 30 sec as before if no switching key responses occurred. There were 10 sessions, each 6 min long. In this phase, the reward outcomes were marbles which could be exchanged for *experimenter-selected* candy during one signal or marbles traded for *self-selected* candy on the other. Again, the colors signaling each outcome were the same for half the subjects in each group and for half they were the opposite.

Findings

Multiple Schedule Performance. The data from the multiple schedule training on Days 1 and 2 gave evidence that the number of people per apartment had no differential effect on response rate. As indicated in Table 2, when reinforcement outcomes alternated automatically, subjects in the two through eight person conditions differed only trivially from one another, $F(6,21) = 1.53$. In all cases, they pressed vigorously when rewarded with immediate candy and considerably less

Table 2. Mean Number of Responses per 30-Sec Period Summed over Trials of Multiple Schedule Procedure

				Marbles traded for	
Number of people living with child	n	Immediate candy	No trade	Self-selected candy	Experimenter-selected candy
2	(3)	138.0	52.3	141.7	134.7
3	(5)	186.2	76.2	176.0	161.2
4	(5)	155.0	45.2	128.0	125.0
5	(4)	176.5	35.0	147.0	152.3
6	(4)	152.0	56.0	133.4	136.2
7	(3)	140.7	34.7	131.3	117.0
8	(4)	165.0	52.0	138.8	131.8

when reinforced with marbles that could not be traded for candy, $t(56) = 5.19$, $p < .001$. Subjects responded only slightly and nonsignificantly more frequently overall for marbles that could be traded for self-selected candy rather than candy selected by the experimenter, $F < 1$. The overall F among multiple schedule conditions was significant, $F(3,63) = 9.47$, $p < .001$; however, there was no significant density $\times$ schedule interaction, $F < 1$.

It is clear that children in all density conditions were about equally capable of learning to respond in a multiple schedule procedure and to distinguish among certain of the outcomes when they were presented successively. The findings from the concurrent schedule procedure, reported in the next section, provide a measure of preference between outcomes when the alternatives were presented simultaneously.[3]

Concurrent Schedule Performance. With this procedure, it will be recalled, a subject could press a button to maintain a given schedule or to switch to the alternative schedule if he so chose. If the subject made no switching response, the schedules were automatically alternated every 30 sec as before. The time spent in each schedule during the 360-sec test period can be taken as a measure of preference. These data were redundant with two other possible dependent variables: total responses in a component and switching responses into a component.

[3]Although we were primarily interested only in responses during the concurrent schedule procedure, preliminary training using the multiple schedule procedure was included for several reasons. This procedure familiarized the subjects with the operation of the apparatus and gave them experience in obtaining the various outcomes. Further, the data provided a baseline for assessing between-groups differences in learning ability *per se*. Finally, the procedure permitted us to assess whether subjects' rate of pressing was different for the various reward outcomes.

Table 3. Time in Seconds (out of a Possible 360) Spent on
Each Component of the Concurrent Schedule

Number of people living with child	Self-selected candy	Experimenter-selected candy
2	274.7	85.3
3	248.8	111.2
4	217.4	142.6
5	234.8	125.2
6	200.0	160.0
7	176.6	183.4
8	180.2	179.8

The introduction of the switching key had a large differential effect
on subjects from the various density conditions. The more crowded a
child's living conditions, the less likely he was, on the average, to use
the switching key. Table 3 shows that children living with eight other
people, for example, spent a mean of 180 sec in each component—
exactly the time allotted to each by the programmed alternation between
schedules. Children from lower density environments used the switching key to spend more time working for marbles that could be traded for
self-selected candy. The correlation between density and time earning
marbles for self-selected candy was $r(27) = -.448, p < .05$.

Conclusions

Children from lower density housing used the choice key more than
their counterparts from high density households, and they specifically
used it to earn self-selected candy rather than experimenter-administered
candy. Why should children living with more people in the same amount
of space be less likely to exercise this kind of control?

First, it is conceivable that children living in conditions of high
density develop deficits in learning ability and thus understood the procedure less well. However, the experimenter always demonstrated each
schedule and made certain that the subject actually pressed the key
several times to see that his responses changed the signal. In addition, a
second experimenter, who knew nothing about the subject's performance during the experiment or his home density, interviewed each
child after the experiment and asked him to explain and demonstrate
every procedure used. All subjects included in the data analyses showed
adequate comprehension by meeting a fixed criterion in response to
these questions. Finally, all subjects except three did learn to respond to

the contingencies of at least the multiple schedule procedure equally well.

Rather than cognitive deficits, perhaps motivational deficits characterize the difference between these subjects. We have suggested that high density living may often undermine one's feelings of choice and control. For example, under highly crowded conditions an individual cannot always have quiet when he wants it, may not be able to sleep when he would like, and so forth. Zlutnick and Altman (1972) have proposed that high density may also engender the "kinds of interpersonal events where people are unable to control their interactions with others and/or where the psychological and physiological costs for interaction are high; and produce subjective feelings of an inability to control interpersonal exchange" (p. 50). Although it is possible to adapt to these conditions, the process may have negative postadaptational consequences. As Glass and Singer (1972) have suggested for uncontrollable noise, this occurs because the individual is subjected to environmental outcomes that he is unable to control by his own responses.

Seligman and Maier (1967) have demonstrated that when organisms learn that their outcomes are independent of any response they make, they no longer learn to perform adaptively when their outcomes are, in fact, contingent on their responses. This phenomenon has been labeled "learned helplessness." If this kind of learning also occurs in some chronically crowded people, it is possible that they may learn to stop trying to control or to make active choices, or they may never learn to exercise effective control or make important choices in the first place.

While the results of the first experiment are consistent with such an interpretation, they very well may have been caused by a variety of other factors. Consequently, a second study was conducted to examine the effects of high density on responses to controllable and uncontrollable outcomes.

Experiment 2

In this study a randomly selected sample of subjects was tested. Subjects were first asked to solve either a solvable or an insolvable problem and then were given a second solvable problem. As Hiroto and Seligman (1975) have pointed out, insolvability in a cognitive task is formally analogous to inescapability, since in both the probability of reinforcement (correct or incorrect or shock or no shock) is independent of responding. It was predicted that control-relevant experiences due to high residential density would influence expectancies formed in an experimental situation that made uncontrollability between response and outcome salient. Specifically, we expected that for all subjects, experience with uncontrollable outcomes would interfere with the acquisition of a subsequent response that could control relevant outcomes, but that crowded children would be more strongly affected by uncontrollability.

The Study

Overview of Experimental Procedure.   The study was designed to examine the interaction of long- and short-term crowding in male and female subjects on a variety of behaviors.[4] Subjects participated in a series of experiments for 3 hr in either crowded or uncrowded same-sex four-person groups. At the conclusion of the experiment each child was interviewed extensively, using both oral and written questionnaires, to establish demographic information relevant to family structure and density.

Subjects.   Using lists provided by all three of the local junior high schools in New Haven, parents of 190 randomly selected seventh and eighth grade students received a letter explaining the study, offering the child payment for his participation, and requesting permission for the child to participate. One hundred seventy-five children received parental permission and three were randomly excluded in order to obtain 43 four-person groups. Of the total sample, 81.5% were black, 14.13% were white, 3.26% were Puerto Rican, and 1.09% were Oriental. Forty-eight percent of the subjects were girls and 52% were boys. All subjects were met at their schools and transported to Yale for the study.

Procedure.   Eight children participated in each time period, assigned four to a room in same-sex groups. The subjects in each room were from different grades and/or different schools and did not know one another. During each period, one four-person group was in a room measuring 8 × 5 and one was in a room measuring 8 × 10 ft but otherwise identical. In each room there were four chairs arranged in a circle. The chairs were spread out, allowing room between subjects, in the uncrowded room, and pushed close together, allowing only cramped leg room, in the crowded area. Before the control-relevant measure, subjects spent 2 hr in these rooms playing games and completing questionnaires. They then were given instructions that explained the experimental task. Whether or not the first problem was solvable, all subjects were told:

---

[4]A full report of this experiment, conducted with John McConahay, is currently in preparation. For present purposes, only those measures relevant to learned helplessness have been discussed here.

> Now I'd like to see whether you can see the pattern in a series of numbers I'm going to show you. The way it works is like this. I have here a list of numbers, all of which are either "0" or "1." I'm going to show you these numbers one at a time, and your job is to say whether the next number is going to be a "0" or "1." Let me show you what I mean, using 3s and 4s for practice.

The experimenter then showed a four-number sequence of 3s and 4s as an example.

> Of course, the pattern will be different in the real game. When we begin, I'd like you to write each answer on a separate page in your booklet. Each time I'll show you what the right answer was.

There were 35 trials. For the solvable set, the pattern of numbers was 0 0 0 1 1. The insolvable pattern was randomly generated. Each four-person group was assigned to a solvable or insolvable condition. Following the first session, all groups received a second set of 35 trials with a solvable pattern of X's and Y's (X Y X X X Y).

Findings

To check the manipulation of response-contingent versus noncontingent learning, the effects of first working on a solvable or an insolvable problem were tested by examining the number correctly identified on the solvable task that followed. Group means were used to assess the experimental effects of pretreatment with solvability and manipulated room size. Sex effects were also tested using these group means for same-sex groups.

As indicated by the means in Table 4, the number correct on the second, solvable problem was greater for groups given the solvable pretreatment than for those previously exposed to insolvability, $F(1,35) = 4.36$, $p < .05$. The analysis of variance revealed no significant main effects of manipulated spatial restriction or sex and no reliable interactions.

The effects of the predispositional variables were tested in a multiple regression analysis using the entire sample. Since the interdependence within groups presumably cut across predispositional variation,

Table 4. Mean Number Correct on Solvable Pattern Two

	Solvability pretreatment	
Short-term crowding	Solvable	Insolvable
Uncrowded	22.875	18.598
Crowded	20.610	18.266

Table 5. Simple Correlation Coefficients for Predispositional Variables:
Solvability, Pretreatment, and Number Correct on Experimental Task

	Sex	Race	Family structure	Parents' education	Density	Solvability
S						
R	−.021					
Fs	.109	−.232[b]				
Pe	−.057	−.579[c]	.258[b]			
D	.042	.311[c]	−.183[a]	−.481[c]		
S	−.006	.012	−.091	.044	.031	
Number correct	−.025	−.247[b]	.120	−.593[c]	.540[c]	.396[c]

[a] $p < .05$.
[b] $p < .01$.
[c] $p < .001$.

this would only reduce rather than increase the multiple correlation. The
predispositional variables tested were sex, race, two- versus one-parent
family, highest education level of either parent (a social class indicator),
and residential density (persons per room). The manipulated variable of
the solvable or insolvable pretreatment was also entered as a main
effect.[5]

The matrix of zero-order correlations is given in Table 5. The data
indicate that race, density, parents' education, and the solvability pretreatment all were significantly related to the number of correct trials on
the second, solvable puzzle. There were also significant relationships
among several of the independent variables, especially between density
and some of the socioeconomic status indicators, such as race and parents' education. The multiple regression analysis appraised the effects of
each independent variable with the others partialed because all the variables were entered simultaneously as a set of main effects.

The results of the multiple regression analysis, given in Table 6,
indicate that the six variables together accounted for 39% of the variance
in number correct on the second puzzle. Parents' education, solvability
of pretreatment, and density had the greatest effects.

It was also predicted that the interaction between residential density
and solvability on the first puzzle would influence ability to learn the

[5]The dichotomous variables were coded as dummy variables as follows: sex (1 = female, 0
= male), race (1 = minority, 0 = white); family structure (1 = two parent, 0 = one parent),
solvability (1 = solvable, 0 = insolvable).

Table 6. Amount of Total Variance in Number Correct on the Second (Solvable) Puzzle Accounted for by Various Independent Variables

Independent variables	Variance accounted for (%)
Results for main effects	
Sex	3
Race	6
Parents' education	10
Two- vs. one-parent family	<1
Residential density	8
Solvable pretreatment	11
$R^2 = .392$, $F(6,165) = 7.53$, $p < .001$.	
Results for main effects and density $\times$ solvability interaction	
Sex	2
Race	6
Parents' education	12
Two- vs. one-parent family	<1
Residential density	7
Solvable pretreatment	5
Density $\times$ solvability	11
$R^2 = .501$, $F(7,164) = 5.19$, $p < .001$.	

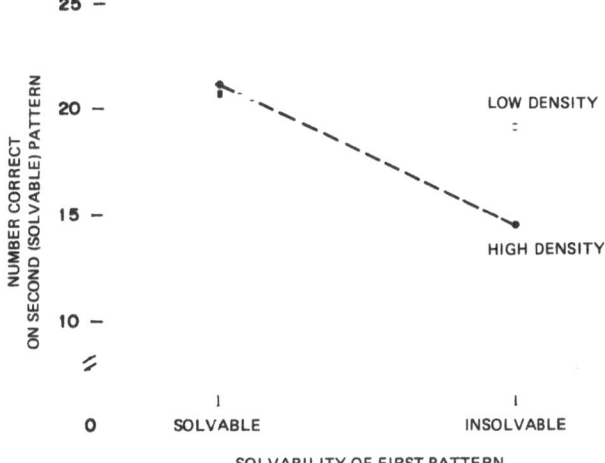

Figure 1. Number correct on a solvable task after pretreatment with either solvable or insolvable problem.

second pattern, which was always solvable. The product of density and solvability was entered as the seventh independent variable in another multiple regression analysis.[6] As indicated in Table 6, its entry increased the multiple correlation by .111, a significant amount, $t(164) = 3.39$, $p < .001$. The means in Figure 1, calculated from the regression equation, show this conditional relationship. When the first puzzle was insolvable, the high density children did more poorly on the next solvable task than did other high-density children who had a solvable puzzle first. Low-density children showed this difference less strongly.

Conclusions

In the present experiment, there was a negative association between number of correct answers on a solvable puzzle and being from a racial minority, parents' education level, residential density, and preexposure to an insolvable puzzle. The interaction of high density and pretreatment with an insolvable puzzle also strongly contributed to the variance in performance on the solvable puzzle. In the animal literature, there is at least one report examining the effects of high density on problem solving. Goeckner, Greenough, and Mead (1973) found that rats reared under conditions of high density showed poorer performance on complex appetitive and avoidance tasks, although there were no significant differences on simple tasks. They used a learned helplessness analysis to explain why many rats, reared in high density, failed even to make a response in a stressful and complex discrimination task. Similarly, we also suggest that, as a consequence of chronic high density living, people may come to feel that they are at the mercy of their environment—unable to control the contingency between their responses and their outcomes.

When the features of the present experiment made salient the noncontingency between responses and outcomes on the puzzle tasks, high-density subjects were slower to solve a second, solvable puzzle. Presumably, this occurred because these subjects came into the laboratory with learning histories that already included extensive helplessness conditioning. Thus, they began working on the first task with expectancies of noncontingency or uncontrollability. Also testing children, Dweck and Reppucci (1973) demonstrated that perception of uncontrollability (as measured by IAR) was strongly related to the deterioration of performance following noncontingent failure. This performance decre-

---

[6]Since the main effects were already in the equation, they were partialed from the product, and, thus, it represents a true interaction term (Cohen & Cohen, 1975, Chap. 8).

ment could be reversed by explicitly training the child to see a contingency between response and failure outcome (Dweck, 1975).

In order to consider the hypothesis regarding expectancies of noncontingency in another way, learning on the first solvable task was examined. Children from high-density housing took significantly longer than low density children to learn the solvable pattern (zero-order correlation $r(84) = -.385$, $p < .001$). Consequently, they did appear to have greater difficulty learning the contingent response. Nonetheless, once given pretraining with a solvable series, they did almost as well on the second solvable task as did low-density children. Having seen that their responses on this type of task were effective, they were better able to learn a similar task. It must be remembered, however, that if they were first given experience with an insolvable problem, they were considerably less able to master the solvable task.[7] This identical pattern of results has recently been reported by Miller (1974) using depressed and nondepressed college students pretreated with escapable or inescapable noise. Miller also suggested that learning to respond to escape the noise forced the depressed group into seeing that something they did really worked.

Both experiments reported here support the view that chronic high density living may reduce one's feelings of choice and control. It is possible that density produces these effects by fostering conditions in which both negative and positive events may be unpredictable and/or uncontrollable. Predictability and perceived control have been shown to often reduce the aversiveness of noxious stimuli (Glass & Singer, 1972; Pervin, 1963; Staub, Tursky, & Schwartz, 1971). Although less well investigated, prediction and control probably have an equally important effect on response to positive outcomes (Wortman, 1975). Adaptation to lack of prediction and control may lead to cognitive and motivational deficits (Glass & Singer, 1972), and expectancies for a lack of control and subsequent helplessness (Wortman & Brehm, 1975). While the present interpretation of the data is still speculative, it is able to deal with and integrate a growing body of literature on the importance of predictability and control for human and animal behavior. However, the particular aspects of the long-term dwelling situation that produced these effects may still be open to question and should be subjected to further investi-

[7]We did not include a no-pretreatment control group because orthogonal manipulations of acute crowding and sex of subjects would have required a substantial increase in the number of subjects tested. More students were simply unavailable to us, within this population, and, thus, on the basis of the present results, it is entirely possible that pretreatment with a solvable task heightened feelings of competence and mastery rather than insolvable puzzle pretreatment enhancing helplessness.

gation. The value of the present studies lies in their emphasis on the possibility of approaching the potential negative effects of chronic density by viewing perceived uncontrollability as a critical mediating process. The impact of this moderator variable should be most evident in those situations that make outcome control and/or choice salient.

References

Aiello, J. R. Epstein, Y. M., & Karlin, R. A., Effects of crowding on electrodermal activity. *Sociological Symposium*, 1975. (a)

Aiello, J. R., Epstein, Y. M., & Karlin, R. A. *Field experimental research on human crowding*. Paper presented at the Eastern Psychological Association meetings, New York, 1975. (b)

Anastasi, A. *Differential psychology*. New York: Macmillan, 1958.

Baron, R. M., Mandel, D. R., Adams, C. A., & Griffen, L. M. *Effects of social density in university residential environments*. Paper presented at the American Psychological Association meetings, Chicago, 1975.

Brigham, T., & Sherman, J. *Effects of choice and immediacy of reinforcement on single response and switching behavior of children*. University of Kansas: Follow-Through Project, 1971.

Calhoun, J. B. Phenomena associated with population density. *Proceedings of the National Academy of Science*, 1961, *47*, 428–449.

Calhoun, J. B. Population density and social pathology. *Scientific American*, 1962, *206*, 139–148.

Cohen, J., & Cohen, P. *Applied multiple regression/correlation analysis for the behavioral sciences*. Hillsdale, New Jersey: Erlbaum, 1975.

Dweck, C. S. The role of expectations and attributions in the alleviation of learned helplessness. *Journal of Personality and Social Psychology*, 1975, *31*, 674–685.

Dweck, C. S., & Reppucci, N. D. Learned helplessness and reinforcement responsibility in children. *Journal of Personality and Social Psychology*, 1973, *25*, 109–116.

Freedman, J. *Crowding and behavior*. San Francisco: Freeman, 1975.

Freedman, J., Heshka, S., & Levy, A. Population density and pathology: Is there a relationship? *Journal of Experimental Social Psychology*, 1975, *11*, 539–552.

Galle, O., Gove, W., & McPherson, J. M. Population density and pathology: What are the relations for man? *Science*, 1972, *176*, 23–30.

Glass, D. C., & Singer, J. E. *Stress and adaptation: Experimental studies of behavioral effects of exposure to aversive events*. New York: Academic Press, 1972.

Goeckner, D. J.. Greenough, W. T., & Mead, W. R. Deficits in learning tasks following chronic overcrowding in rats. *Journal of Personality and Social Psychology*, 1973, *28*, 256–261.

Hiroto, D. S., & Seligman, M. E. P. Generality of learned helplessness in man. *Journal of Personality and Social Psychology*, 1975, *31*, 311–327.

Hutt, C., & Vaizey, M. J. Differential effects of group density on social behavior. *Nature* (*London*), 1966, *209*, 1371–1372.

Miller, W. R. *Learned helplessness in depressed and nondepressed students*. Unpublished doctoral dissertation, University of Pennsylvania, 1974.

Pervin, L. A. The need to predict and control under conditions of threat. *Journal of Personality*, 1963, *31*, 570–587.

Proshansky, H. M., Ittelson, W., & Rivlin, L. G. Freedom of choice and behavior in a physical setting. In H. Proshansky, W. Ittelson, & L. Rivlin (Eds.), *Environmental psychology*. New York: Holt, 1970, 173–183.

Saegert, S., MacKintosh, E., & West, S. Two studies of crowding in urban public spaces. *Environment and Behavior*, 1975, 7, 159–184.

Schmitt, R. C. Density, delinquency, and crime in Honolulu. *Sociology and Social Research*, 1957, 41, 274–276.

Schmitt, R. C. Density, health, and social disorganization. *Journal of the American Institute of Planners*, 1966, 32, 38–40.

Seligman, M. E. P., & Maier, S. Failure to escape traumatic shock. *Journal of Experimental Psychology*, 1967, 74, 1–9.

Smith, S., & Haythorn, W. W. Effects of compatability, crowding, group size, and leadership seniority on stress, anxiety, hostility, and annoyance in isolated groups. *Journal of Personality and Social Psychology*, 1972, 22, 67–69.

Staub, E., Tursky, B., & Schwartz, G. Self-control and predictability: Their effect on reactions to aversive stimulation. *Journal of Personality and Social Psychology*, 1971, 18. 157–162.

Winsborough, H. H. The social consequences of high population density. *Law and Contemporary Problems*, 1965, 30, 120–126.

Wortman, C. Some determinants of perceived control. *Journal of Personality and Social Psychology*, 1975, 31, 282–294.

Wortman, C., & Brehm, J. Responses to uncontrollable outcomes. In L. Berkowitz (Ed.), *Advances in experimental social psychology*, 1975, 8, 278–336.

Zlutnick, S., & Altman, I. Crowding and human behavior. In J. F. Wohlwill & D. H. Carson (Eds.), *Environment and the social sciences*. Washington, D.C.: American Psychological Association, 1972.

6

Field Research on the Effects of Crowding in Prisons and on Offshore Drilling Platforms

Verne C. Cox, Paul B. Paulus, Garvin McCain, and Janette K. Schkade

Much of the contemporary interest in the possible psychological effects of crowding[1] stems from the combined influence of Calhoun's (1962) classic studies on crowding effects in animals and the concern about potential psychological consequences of human population growth. Calhoun's work raised the possibility that humans might display behavior pathology similar to that observed in severely crowded rats. Increasing population growth offers the possibility that the degree of crowding that yielded behavior pathology in Calhoun's studies may be approached or achieved by large segments of the world's human population. Even now it is commonly assumed that existing degrees of crowding in urban areas contribute substantially to behavior pathology (Zlutnick & Altman, 1972).

[1]The use of terms such as *crowding* has not been consistent among authors. Following Rapoport (1975), Stokols (1972), and others, we will use the term *spatial density* to refer to the number of square feet of floor space in a particular unit. *Social density* will refer to the number of individuals in a housing unit. *Perceived social density* will refer to the subjective evaluation of social density by occupants of a unit. *Crowding* will refer to a condition of either social or spatial density experienced as aversive by the occupants of a unit.

Verne C. Cox, Paul B. Paulus, Garvin McCain, and Janette K. Schkade • Department of Psychology, University of Texas at Arlington, Arlington, Texas 76019. The research presented was supported by a grant from the Organized Research Fund of The University of Texas at Arlington.

While these concerns are legitimate, it has been difficult to determine their validity. First, crowding in natural settings is typically confounded with a multitude of other variables. Second, most human crowding research has focused on social and spatial density conditions that differ substantially in severity and duration from conditions analogous to the crowding employed in Calhoun's animal work. Consequently, there are few data available regarding human reactions to long-term, intense, and inescapable crowding. In this chapter we will discuss the advantages of examining this question in prisons, jails, and offshore drilling platforms, as well as other field settings. Our past research in prisons and jails will be reviewed, and our observations from our initial visits to offshore drilling platforms will be discussed.

There are three basic approaches to examining human reactions to crowding: laboratory studies where intense crowding can be achieved with experimental control, statistical analyses of correlates of crowding in large social units such as areas of cities (e.g., census tracts), and field research in specific settings reflecting various ranges of crowding. Each approach has advantages and limitations. Laboratory investigations typically involve volunteer subjects, often paid for participation, who are usually exposed to no more than several hours of various degrees of crowding. While maximal space limitation can be achieved in these settings, practical and ethical considerations limit their duration, and the ability to voluntarily withdraw from the experimental conditions probably affects tolerance for intense crowding in the laboratory. Many such studies find little or no effect on a variety of dependent variables. The work of Freedman and his co-workers is a good example of this type of research. These investigators found, in their initial studies, no positive or negative effects of short-term intense crowding. (Freedman, Klevansky, & Ehrlich, 1971). Subsequent studies reported sex-related positive and negative effects (e.g., Epstein & Karlin, 1975; Freedman, Levy, Buchanan, & Price, 1972; Ross, Layton, Erickson, & Schopler, 1973) of laboratory-induced crowding on affect and social behavior. None of the negative effects reported in these studies could be considered pathological. Recent research reported by Paulus and his co-workers has indicated that laboratory-induced crowding can have deleterious effects. One study found that decreased room size, increased group size, and increased proximity all led to decrements in task performance (Paulus, Annis, Seta, Schkade, & Matthews, 1976). These effects were found with both males and females. A second study suggested that these results are limited to situations in which others are seen as potentially negative or aversive stimuli (Seta, Paulus, & Schkade, 1976). Even laboratory effects such as those described above may not relate to crowding in natural environs where exposure can be prolonged and sometimes inescapable.

To circumvent problems posed by laboratory investigations of crowding, unfortunately at the cost of introducing others, some investigators have applied statistical analyses to relationships between various degrees of crowding in cities and various indices of pathology such as mental illness, suicide, crime, and illness. These types of studies have consistently yielded conflicting results. For example, a recent report by Galle, Gove, and McPherson (1972) reported significant relationships between measures of crowding and several categories of social pathology, such as juvenile delinquency and mental hospital admissions. However, their findings are rather weak and their statistical techniques questionable (cf. Freedman, Heshka, & Levy, 1975). Freedman *et al.* (1975), utilizing various areas of New York City, found practically no relationship between measures of crowding and pathology once the effects of ethnicity and income level were taken into account. Problems associated with correlational analyses of demographic data are too well known to need comment. The best one can conclude from studies such as Galle *et al.* (1972) and Freedman *et al.* (1975) is that statistical surveys do not make a clear case either for or against the existence of disruptive influences induced by crowding. Quite apart from methodological problems involved in statistical investigations of crowding in cities, these environments do not ordinarily represent, at least for the present, prolonged, intense, inescapable exposure to crowding. The residents of cities have a great deal of control over their exposure to crowded situations. They can stay in their homes or venture out only during times of minimum social density. If their housing unit is crowded they can venture outside at will. Furthermore, crowding within the home primarily involves dealing with family members and not strangers or potentially aversive others. Even at rush hours in cities, intense crowding usually involves only a small part of the day. Additionally, a recent study suggests that the levels of social and nonsocial stimulation in cities and towns of moderate size may not be substantially different (Korte, Ypma, & Toppen, 1975). Thus, cities at present do not appear to represent situations analogous to Calhoun's type of crowded situations.

The remaining basic approach to the study of crowding is the use of specific field settings such as dormitories (e.g., Aiello, Epstein, & Karlin, 1975; Baron, Mandel, Adams, & Griffen, 1976; Valins & Baum, 1973), naval vessels (Dean, Pugh, & Gunderson, 1975), housing projects (McCarthy & Saegert, Chapter 4, this volume), a train station (Mac-Kintosh, West, & Saegert, 1975), and stores (Langer & Saegert, 1977; MacKintosh *et al.*, 1975). While these settings in most cases involve moderate degrees of crowding that do not approach conditions analogous to Calhoun's conditions, they have yielded results that indicate that crowding can have negative psychological effects. For example, it has been found that crowding produces negative affect and social withdrawal in dorms and housing projects (Aiello *et al.*, 1975; Baron *et al.*, 1976; McCarthy & Saegert, Chapter 4; Valins & Baum, 1973), increased illness complaints on naval ships (Dean *et al.*, 1975), and disruption of cognitive functioning in stores and train stations (Langer & Saegert, 1977; MacKintosh *et al.*, 1975).

Our particular interests for the past 6 years have focused on field settings that provide variations in degree of crowding that include what most would agree is intense and prolonged crowding. We believe that studies of the effects of intense crowding will provide information regarding how humans might respond to situations analogous to those that yield pathology in animal studies. Such studies may also identify variables that may be operating in a less obvious fashion at the lesser degrees of crowding that are more commonly experienced in contemporary societies.

Prisons and Jail Settings

Field settings that provide intense levels of crowding are not common, and fewer still are accessible for research purposes. After considering a number of possibilities, we decided that prisons and jails might provide an environment for studying the effects of prolonged, inescapable, and intense crowding (Paulus, McCain, & Cox, 1973). With the aid of grant support from the Hogg Foundation we spent several months visiting a number of federal and state prison sites and local jails to determine the feasibility of conducting crowding research in these environments. We were particularly interested in finding accessible facilities where architectural features provided a wide range of cell and dormitory sizes and population levels. We wanted to minimize the need for cross-institutional comparisons so that all our subjects would be exposed to the same administrative milieu. Of the 11 institutions visited we found 2 that met our criteria. Tables 1 and 2 illustrate the variation we found in social and spatial density in a federal prison (Texarkana Federal Correctional Institution, TFCI) and a nearby jail (Dallas County Jail, DCJ). Data we have collected from these facilities, over several years, have convinced us that prolonged exposure to high social density can produce a variety of effects that are not revealed in data collected from laboratory studies and statistical surveys.

At TFCI housing ranged from 1-man units to 46-man dormitories.

Table 1. Cell Arrangements: Dallas County Jail

Number of occupants	Length	Width	Square feet per man
Examples of cell arrays that allow examination of variations in spatial density with social density held constant			
8	15 ft. 11 in.	8 ft.	15
8	13 ft.	6 ft. 6 in.	10
4	13 ft.	5 ft. 6 in.	18
4	17 ft.	8 ft.	34
Examples of cell arrays that allow examination of variations in social density with spatial density held constant			
8	15 ft. 1 in.	8 ft.	15
6	13 ft.	6 ft. 8 in.	15
8	13 ft.	6 ft. 6 in.	10
12	15 ft. 11 in.	8 ft.	10

Spatial density during our visits ranged from 21 to 84 square feet per man. Single cells typically provided 55 square feet per man. The residents are of course there involuntarily and have little control over their housing assignment. Although the inmates are not confined to their housing unit at all times, they must be there from 10:00 P.M. until 7:00 A.M. and at several times during the day for roll call. There is a high turnover among inmates within various housing units. Dallas County Jail at the time of our visits involved much more intense crowding. The units ranged from 1-man cells to 70-man dorms. Spatial density during our visits ranged from 10 to 174 square feet per man. Single-man cells typically provided 65 square feet per man. Inmates, unlike in the prison setting, were confined to their cells 24 hours a day. High turnover and

Table 2. Texarkana Federal Correctional Institution

Example	Number of occupants	Square feet per man
Example of variations in population concentration with social and spatial density confounded in the first example and with spatial density constant and social density varied in the second example		
I	44	30.7
	28	49.6
II	33	83.6
	21	84.3

indeterminacy of stay are also characteristic of this jail. To put these figures in perspective it should be realized that 350 square feet of living space per person is considered desirable in American homes (Altman, 1975, p. 177). A 1,000-square-foot house or apartment would have to have approximately 50 residents to approximate the spatial and social density frequently experienced at this facility at the time of our visits. Thus, prisons can provide spatial and social density values that can constitute crowded conditions. The inescapability of the crowding in prisons and the lack of control over housing assignment by the inmates may also be important stress factors. Work by Glass and Singer (1972) and Sherrod (1974) suggests that to the extent that an individual can escape or control his exposure to stressors such as noise and crowding, the effects of the stressor may be attenuated.

Our first studies focused on the effects of social and spatial density on an individual's criterion of overcrowding (Paulus, Cox, McCain, & Chandler, 1975). We employed a variation of a task devised by Desor (1972). Inmates at TFCI were asked to place human figurines in a model room until it appeared overcrowded. This task provides a means for determination of individual differences in thresholds for perception of specific combinations of social and spatial densities as crowded. We also obtained housing histories that indicated the type of housing and length of stay in each type of housing prior to and including the housing of inmates at the time of our visits. The Desor task scores indicated that high conditions of density were related to a lower criterion for the number of people that constituted crowding. Of particular interest were our results that indicated that this tolerance for crowding represented by those high density conditions declined as a function of length of time in such conditions, whereas no relationship was found for the inmates in single-cell housing. Each inmate was also administered five semantic differential scales assessing the affective reactions to housing units. Increases in social density, but not spatial density, were related to increased negative affect among the inmates. Furthermore, decreased tolerance of crowding was found to be related to negative evaluation of physical environs. Dean *et al.* (1975) have reported a similar finding for crew members of naval vessels.

These results suggest that high social density in prisons can produce negative emotional responses and a lessened tolerance for crowding. The fact that tolerance for crowding decreased over time is counter to notions that man can easily adapt to crowded situations, but consistent with recent models that suggest that crowding will lead to increased valuation of low levels of social density (Altman, 1975; Stokols, 1972).

Our subsequent studies have investigated the possibility that high social density as encountered in prisons and jails would generate indications of psychological stress. We chose two measures that were feasible to employ in prison and jail environs and have been interpreted as expressions of psychological stress in other contexts. These were illness complaint rates (McCain, Cox, & Paulus, 1976) and palmar sweat. Hinkle (1961) and more recently Kiritz and Moos (1974) have argued that illnesses are encouraged by psychological stress generated by certain social environments. Physiological mechanisms by which such effects might be mediated have recently been elaborated by Stein, Schiani, & Camerino, (1976). Illness complaint rates have been shown to be systematically related to social and spatial density in naval vessels (Dean *et al.*, 1975) and college dormitories (Jacobs, Spilken, & Norman, 1969). In our investigations of the relationship between crowding and illness complaints we excluded complaints of flu and colds, though even these illnesses are not facilitated by crowding-induced contagion to the degree many assume (Dean *et al.*, 1975).

In the prison setting we were able to compare medical records and housing history for 64 inmates that had lived in the same housing for at least 30 days prior to our visit. In the county jail we examined "kites," which are notes indicating a desire to see a nurse or physician. We calculated, for a 5-week period, the frequency of kites generated by units of high spatial and social density as compared to low density housing units. The most frequent complaints in our samples were headache, nausea, rash, sinus, constipation, chest pain, and asthma. We found that in both TFCI and DCJ illness complaint rates were approximately twice as high in more crowded conditions. These results are consistent with those reported by Dean *et al.* (1975) for naval vessels. Illness complaint rate has been shown to be determined, in part, by psychological stress, and this suggests at least one important deleterious effect that may vary as a function of degree of crowding.

Recently we have examined the possibility that another expression of social stress, palmar sweat, may vary with crowding. This relatively simple measure has been employed as an index of arousal (cf. Dabbs, Johnson, & Leventhal, 1968) and has been shown to vary with social stress generated by audiences and group competition in laboratory studies (Cohen & Davis, 1973; Martens, 1969). We were able to examine the differential contributions of both spatial and social density to this index of stress.

Palmar sweat prints were obtained from 46 inmate volunteers at the Texarkana Federal Correctional Institution. These inmates were assembled in groups of four and their prints were taken simultaneously. These prints are obtained by dabbing the finger with a graphite-based solution that dries in about 20 seconds. The resulting print is then removed with transparent tape and mounted on a slide. Open pores appear as holes in the print. The print is then projected using a microprojector and the number of open pores in a 4-mm-square area around the central whorl are counted by two trained judges. The subject's palmar sweat score was obtained by averaging the scores of the two judges. The scores of the two judges were highly correlated ($r = +.98$). The palmar sweat scores were examined for correlation with the inmate's social density and spatial density. A positive correlation of palmar sweating with social density was obtained ($r = .35$, $df = 44$, $p < .02$). The relationship between spatial density and palmar sweating was negligible ($r = +.06$). Since inmates in units of low social density were found to have been in the prison longer, the social density effect could have been a function of the length of time in prison instead of the type of housing unit. However, the correlation of total days in prison for an inmate and the degree of palmar sweating was not significant ($r = -.08$). These results suggest that high levels of crowding in prisons can increase physiological expressions of stress. These results are also consistent with D'Atri's (1975) finding that crowding in prisons can affect another physiological measure, blood pressure. Our findings suggest that D'Atri's results were probably mediated by social density but not spatial density.

Offshore Oil-Drilling Platforms

Our interest in examining the consequences of intense chronic crowding in field settings has led us to seek other nonprison environments that expose humans to intense crowding. One possibility we have begun to explore is the offshore oil-drilling platform. These facilities are often situated 20 to 75 miles offshore and provide a wide range of social and spatial density. We recently visited two platforms and lived on the platforms for 2 days. Our initial visits were intended to determine the suitability of the platforms for crowding research. However, we did conduct a number of interviews that provided preliminary observations regarding attitudes of workers toward spatial and social density on the platforms.

In domestic waters, platform workers have a 7-day-on and 7-day-off work cycle, and in foreign waters the work cycle is typically 28 days on and 28 days off. Work shifts are typically 12 hours per day. Social and spatial density on some platforms is consistently high and on other platforms periodically reaches high values. Housing quarters can range from 1- and 2-man rooms to 6-man rooms and dormitories. Some platforms experience periodic increases in social and spatial density as additional technical service people visit during particular phases in drilling. On these platforms a consistent core group of workers will periodically experience an increase in social and spatial density, providing the possibility of longitudinal study of the effects of variations in social and spatial density. In addition, the platforms vary widely in the amount of social and spatial density that typically prevails. Thus far we have found that social and spatial density can reach intense levels and there can be strong negative reactions to such crowding. On Platform 1, space per man during our 2-day stay was approximately 11 square feet in bunking quarters occupied by six men. We were housed in these conditions and can report that it is quite unpleasant. We visited this particular rig during a phase in drilling when the normal crew was complemented by additional technical personnel. More typically, when technical personnel are not present, 6-man quarters on this platform are occupied by 4 men. On this platform there was little opportunity to withdraw from social stimulation because the only available nonwork space was a small dining and recreation area that was constantly occupied by several crew members. The total population of this platform at the time of our visit was 46, which is full capacity, but more typically the platform crew is around 35. On Platform 2, the social and spatial density was in general much lower, though at least one 6-man cabin was at full occupancy. The capacity of this rig is 74 men, but at the time of our visit the crew population was 34. Housing units were more spacious than on Platform 1, with 4-man cabins providing 28.7 square feet per man and 6-man cabins 31.5 square feet per man. At the time of our visits the actual number of occupants in 4- and 6-man cabins ranged from 2 to 6. The dining area was large and several recreation rooms were available.

Informal observations and interviews revealed some interesting differences between the workers on the two platforms. The men on the crowded platform, Platform 1, were reluctant to be interviewed. They seemed generally unhappy about life on the rig and frequent complaints were voiced about the 6-man cabins. Two- to 4-man units were preferred and of equal desirability. However, 6-man units were viewed as qualitatively different and highly undesirable. We intend to explore the possibility that these attitudes reflect critical thresholds of social density that elicit negative reactions. The men on the socially and spatially less dense Platform 2 were quite satisfied with their living quarters. They were interested in being interviewed and these interviews indicated very positive feelings about their fellow workers and their environment. While other differences between the platforms may contribute to the observed differences, the differential reactions to being interviewed and the reactions to the 6-man units would seem quite consistent with recent notions of crowding (Altman, 1975: Stokols, 1972) that suggest that high levels of crowding should lead to attempts to reduce the level of social interactions.

On Platform 2 we conducted questionnaire interviews with 20 of the crew members and focused on their attitudes toward particular spatial and social density combinations. None of the crew members, including those housed in 6-man cabins, were willing to sacrifice space per person in order to have fewer occupants. It will be recalled that on Platform 1, where space per man in 6-man cabins was much less than on Platform 2, 6 occupants per cabin was considered very undesirable. None of our questions reliably differentiated between men living in cabins of 1–3 men versus 4–6 men. It may well be that the negative effects of various degrees of social density are evident only above particular minimum values of spatial density.

We were unable to visit Platform 3 due to high seas but were able to obtain blueprints from which to determine the housing arrangements. Platform 3 is a very large platform that can accommodate over 100 crew members, and it has living arrangements with spatial density values of 32 and 33 square feet per man when the 4- and 6-man cabins are filled to capacity. We have also learned of other platform living arrangements that involve dormitory facilities with combined eating, toilet, and sleeping facilities for approximately 30 men in very limited space.

Our initial visits were primarily oriented toward determining the nature of these environments and the feasibility of research. Our future research on offshore drilling platforms should provide more extensive data on the psychological effects of various degrees of social and spatial density on drilling platforms.

ACKNOWLEDGMENTS

Thanks are due Angela Annis, Jane Chandler, John Seta, Margaret Short, Dianne Ross, and Paul Watson, who assisted in data collection and analysis. We are grateful to J. C. Craft, Donald Gaddy, and Bob Thomas of Penrod Drilling Company. Work on the drilling rigs would have been impossible without their generous aid and advice. We also wish to thank the employees and inmates of the Federal Bureau of Prisons and Dallas County Jail, Sheriff Clarence Jones, and James Kitching for their cooperation and help.

References

Aiello, J. R., Epstein, Y. M., & Karlin, R. A. *Field experimental research on human crowding.* Presented at the meeting of the Eastern Psychological Association, New York City, 1975.

Altman, I. *The environment and social behavior: Privacy, personal space, territory and crowding.* Monterey, California: Brooks/Cole, 1975.

Baron, R. M., Mandel, D. R., Adams, C. A., & Griffen, L. M. Effects of social density in university residential environments. *Journal of Personality and Social Psychology*, 1976, *34*, 434-446.

Calhoun, J. B. Population density and social pathology. *Scientific American*, 1962, *206*, 139-148.

Cohen, J., & Davis, J. Effects of audience status, evaluation, and time of action on performance with hidden-word problems. *Journal of Personality and Social Psychology*, 1973, *27*, 74-85.

Dabbs, J. M., Jr., Johnson, J. E., & Levanthal, H. Palmar sweating: A quick and simple measure. *Journal of Experimental Psychology*, 1968, *78*, 347-350.

D'Atri, D. A. Psychophysiological responses to crowding. *Environment and Behavior*, 1975, *7*, 237-252.

Dean, L. M., Pugh, W. M., & Gunderson, E. K. E. Spatial and perceptual components of crowding: Effects on health and satisfaction. *Environment and Behavior*, 1975, *7*, 225-236.

Desor, J. A. Toward a psychological theory of crowding. *Journal of Personality and Social Psychology*, 1972, *21*, 79-83.

Epstein, Y. M., & Karlin, R. A. Effects of acute experimental crowding. *Journal of Applied Social Psychology*, 1975, *5*, 34-53.

Freedman, J. L., Heshka, S., & Levy, A. Population density and pathology: Is there a relationship? *Journal of Experimental Social Psychology*, 1975, *11*, 539-552.

Freedman, J. L., Klevansky, S., & Ehrlich, P. R. The effect of crowding on human task performance. *Journal of Applied Social Psychology*, 1971, *1*, 7-25.

Freedman, J. L., Levy, A. S., Buchanan, R. W., & Price, J. Crowding and human aggressiveness. *Journal of Experimental Social Psychology*, 1972, *8*, 528-548.

Galle, O. R., Gove, W. R., & McPherson, J. M. Population density and pathology: What are the relationships for man? *Science*, 1972, *176*, 23-30.

Glass, D. C., & Singer, J. E. *Urban stress: Experiments on noise and social stressors*. New York: Academic Press, 1972.

Hinkle, L. E. Ecological observations of the relation of physical illness, mental illness, and the social environment. *Psychosomatic Medicine*, 1961, *23*, 289-297.

Jacobs, M. A., Spilken, A., & Norman, M. Relationship of life change, maladaptive aggression, and upper respiratory infection in male college students. *Psychosomatic Medicine*, 1969, *31*, 31-44.

Kiritz, S., & Moos, R. H. Physiological effects of social environments. *Psychosomatic Medicine*, 1974, *36*, 96-114.

Korte, C., Ypma, I., & Toppen, A. Helpfulness in Dutch society as a function of urbanization and environmental input level. *Journal of Personality and Social Psychology*, 1975, *32*, 996-1003.

Langer, E., & Saegert, S. Crowding and cognitive control. *Journal of Personality and Social Psychology*, 1977, *35*, 175-182.

MacKintosh, E., West, S., & Saegert, S. Two studies of crowding in urban public spaces. *Environment and Behavior*, 1975, *7*, 159-184.

Martens, R. Palmer sweating and the presence of an audience. *Journal of Experimental Social Psychology*, 1969, *5*, 371-374.

McCain, G., Cox, V. C., & Paulus, P. B. The relationship between illness complaints and degree of crowding in a prison environment. *Environment and Behavior*, 1976, *8*, 283-290.

McCarthy, D. P., & Saegert, S. Residential density, social overload, and social withdrawal. In J. R. Aiello & A. Baum (Eds.), *Residential crowding and design*. New York: Plenum, 1979.

Paulus, P. B., Annis, A. B., Seta, J. J., Schkade, J. K., & Matthews, R. W. Density does affect task performance. *Journal of Personality and Social Psychology*, 1976, *34*, 248-253.

Paulus, P. B., Cox, V. C., McCain, G., & Chandler, J. Some effects of crowding in a prison environment. *Journal of Applied Social Psychology*, 1975, 5, 86–91.

Paulus, P. B., McCain, G., & Cox, V. C. A note on the use of prisons as environments for investigation of crowding. *Bulletin of the Psychonomic Society*, 1973, 1(6A), 427–428.

Rapoport, A. Toward a redefinition of density. *Environment and Behavior*, 1975, 7, 133–158.

Ross, M., Layton, B., Erickson, B. M., & Schopler, J. Affect, facial regard, and reactions to crowding. *Journal of Personality and Social Psychology*, 1973, 28, 69–76.

Seta, J. J., Paulus, P. B., & Schkade, J. K. The effects of group size and proximity under competitive and cooperative conditions. *Journal of Personality and Social Psychology*, 1976, 34, 47–53.

Sherrod, D. R. Crowding, perceived control, and behavioral aftereffects. *Journal of Applied Social Psychology*, 1974, 4, 171–186.

Stein, M., Schiani, R. C., & Camerino, M. The influence of brain and behavior on the immune system. *Science*, 1976, 191, 435–440.

Stokols, D. A social-psychological model of human crowding. *Journal of American Institute of Planners*, 1972, 38, 72–83.

Valins, S., & Baum, A. Residential group size, social interaction, and crowding. *Environment and Behavior*, 1973, 5, 421–439.

Zlutnick, S., & Altman, I. Crowding and human behavior. In J. F. Wohlwill & D. H. Carson (Eds.), *Environment and the social sciences: Perspectives and applications.* Washington, D.C.: American Psychological Association, 1972.

7

Perception of Residential Crowding, Classroom Experiences, and Student Health*

Daniel Stokols, Walter Ohlig, and Susan M. Resnick

Introduction

A core concern of environmental design research is the impact of the physical and social environment on human health and behavior. Professional designers often approach this issue through direct observation of the relationships between architectural variables and behavioral patterns. Design-oriented behavioral scientists, while sharing designers' concern with the direct linkages between objective environments and overt behaviors, more commonly approach the environment–behavior interface by way of theoretical, or intervening constructs. These constructs help to specify the social and physical conditions under which

*Portions of this chapter were presented in a symposium on Design for Communality and Privacy at the Environmental Design Research Association, Lawrence, Kansas, April 1975, and in a symposium on Habitability in Unusual Environments at the Annual Conference of the American Psychological Association, Chicago, Illinois, August 1975. This paper was previously published in *Human Ecology: An Interdisciplinary Journal*, Vol. 6, No. 3, pp. 233–252. Copyright 1978 by Plenum Publishing Corporation.

Daniel Stokols, Walter Ohlig, and Susan M. Resnick • Program in Social Ecology, University of California, Irvine, California 92717.

specific features of the physical environment might correlate with certain types of behavior.

The design-practitioner and behavioral-scientist perspectives are, of course, complementary in the sense that the first keeps environmental-design research firmly tied to concrete, measurable variables, while the second assists the researcher in predicting, *a priori*, those environmental dimensions that will exert the greatest influence on behavior across diverse situations.

The utility of combining design and behavioral perspectives can be illustrated through a consideration of the issue of human crowding. An architecturally oriented analysis of crowding would focus primarily on physical variables, such as the amount and arrangement of space, and the correlations between these variables and behavioral patterns within specific settings. A typical assumption underlying this approach is that spatial limitation is associated with a number of negative behavioral effects, including withdrawal from social interaction, impairment of task performance, and pathological behavior. The problem with a purely physicalistic perspective on crowding is that it does not account for numerous situations in which limited space is associated with positive rather than negative behavioral consequences; for example, work situations in which the proximity of others enhances collective task performance; or crowded parties that promote camaraderie among members of a group. In order to predict where spatial limitation will induce physiological, psychological, and behavioral problems, it becomes necessary to consider the social-psychological dimensions of various situations.

A behavioral science approach to crowding attempts to identify social-psychological factors that mediate the impact of architectural (especially spatial) variables on behavior. A central assumption underlying this approach is that the categorization of environments in terms of both physical and social-psychological dimensions should provide a basis for developing design guidelines pertaining to several related questions, such as: In what types of environments will spatial limitation lead to major disruptions in individual and interpersonal activities? What kinds of adaptive strategies are available to occupants of high-density settings? To what extent will psychological and behavioral deficits associated with crowding in one setting generalize to other situations?

These questions are examined below in relation to a typology of crowding experiences (Stokols, 1976, 1978). The typology focuses on the subjective experience of crowding rather than on conditions of high density that may or may not be related to perceived crowding. A number of derivative hypotheses pertaining to the intensity, persistence, and reducibility of crowding experiences are discussed, and some preliminary research on the generalizability of crowding experiences from one situation to another is presented. It is assumed throughout this discussion that a refinement of the crowding construct will provide a basis for predicting crowding potentials and related behavioral impairments as a function of both physical and social dimensions of the environment, and that such prediction ultimately will contribute to the design of physical settings that are maximally congruent with the needs of their users.

A Typology of Crowding Experiences

The typology is based on the distinction between density, a physical condition of limited space, and crowding, a subjective experience of psychological stress in which one's demand for space exceeds the available supply (Stokols, 1972a). A basic assumption relating to this distinction is that increased demand for space can arise not only in response to direct spatial restriction but also as a result of social circumstances that sensitize the individual to potential problems posed by continued proximity with others.

Recent analyses have suggested a variety of nonspatial antecedents of crowding. Stimulus overload models, for example, posit that the experience of crowding is heightened by excessive social stimulation (cf. Desor, 1972; Esser, 1972; Milgram, 1970; Valins & Baum, 1973; Zlutnick & Altman, 1972). Behavioral constraint formulations link the perception of crowding to restraints on behavioral freedom and infringements on privacy imposed by the proximity of others (cf. Proshansky, Ittelson, & Rivlin, 1970; Stokols, 1972b). And ecological perspectives on crowding suggest that increased demand for space may result from a scarcity of social and/or physical resources in the setting (cf. Hanson & Wicker, 1973; Wicker, 1973).

Stimulus overload, behavioral constraint, and ecological theories of crowding converge on the assumptions that crowding (1) involves the perception of insufficient control over the environment, and (2) increases the desire to put more space between oneself and others as a means of avoiding actual or anticipated interferences. The shared utility of these analyses is that they provide insights into the nature and determinants of perceived crowding. Their major limitation is that they offer few clues concerning the parameters of crowding intensity and persistence. The conditions under which overstimulation, behavioral constraints, and resource scarcities lead to the most disruptive experiences of crowding remain unspecified.

An additional assumption concerning the nature of crowding is required to permit an identification of factors that mediate the intensity and persistence of crowding experiences, namely: Increased demand for space will be most intense, persistent, and difficult to resolve when it is associated with perceived threats to physical or psychological security. Proximity with dangerous or insulting persons, for example, would lead to more intense crowding than the same degree of proximity with others who are seen as posing no threat to the individual's security.

The present typology of crowding experiences incorporates two dimensions that help to "sort out" the determinants of crowding intensity and persistence: neutral-personal thwartings and primary-secondary environments. The thwarting dimension pertains to the nature of interferences imposed by proximity with others. Neutral thwartings are essentially unintentional annoyances stemming from either the social or the nonsocial environment, whereas personal thwartings are those interferences intentionally imposed on the individual by other persons (cf. Stokols, 1975). Under conditions of *neutral crowding*, the need for more space relates primarily to physical concerns, such as the restriction of movement and the discomforts associated with high-density conditions. To escape these inconveniences, the individual desires more control over the physical, i.e., spatial, environment. In situations of *personal crowding*, increased demand for space relates to both physical and social concerns. Here, the salience of physical inconveniences is increased by the presence of hostile or unpredictable others. To resolve feelings of crowding, the individual must gain control over social as well as physical aspects of the environment.

The primary-secondary dimension of the model concerns the continuity of social encounters in a particular setting, the psychological centrality of behavioral functions performed within the setting, and the degree to which social relations occur on a personal or anonymous level. *Primary environments* (e.g., residential and work environments) are those in which an individual spends much time, relates to others on a personal basis, and engages in personally important activities. *Secondary environments* (e.g., transportation and commercial environments) are those in which the individual's encounters with others are relatively transitory, anonymous, and inconsequential.

The model of crowding experiences outlined above suggests several hypotheses. First, it can be predicted that experiences of personal crowding will be more intense, persistent, and difficult to resolve than those of neutral crowding, since the former are more likely to involve perceived threats to one's physical safety or self-esteem and to induce frustration of expectancies regarding the adequacy of space. Furthermore, assuming that the individual is confined to the situation, perceptual and cognitive modes of adaptation to crowding (e.g., ignoring spatial constraints, adopting more favorable attitudes toward other occupants of the area) will be less viable in situations of personal crowding, because the potential for social conflict is greater there than in instances of neutral crowding.

A second hypothesis is that crowding experiences will be of greater intensity and duration in primary environments than in secondary ones. This prediction is based on the assumption that primary settings tend to be associated with higher expectations of personal control along a greater dive ,ity of need dimensions and, therefore, proximity-related interferences will be more likely to thwart personally important needs and goals in primary vis-à-vis secondary environments.

A third hypothesis pertains to the generalizability of crowding experiences from one setting to another. It is expected that personal crowding experiences, particularly in the context of a primary environment, will generalize more readily to other situations than will neutral crowding experiences. (The generalizability of crowding experiences could be measured, for example, in terms of a person's need for space, susceptibility to performance deficits, and medical complaints across a variety of settings, subsequent to his or her experience of crowding within a particular situation.) The main assumption underlying this prediction is that personal crowding, because it typically involves ambivalent or negative attitudes toward others, provides a cognitive base from which anxieties about proximity with certain persons in the setting can generalize to other people in different situations.[1] In contrast, the impact of neutral crowding experiences, which are less closely associated with persisting attitudinal changes, would be more confined to the immediate situation.

The present research provides preliminary data that pertain directly to the third hypothesis and are of general relevance to the first and second predictions mentioned above. The data were gathered through a three-part campus survey in which college students evaluated their residential environments (e.g., dormitory suites, apartments) during the second and third weeks of the academic quarter, and rated the amenity of a particular classroom setting during the fourth week of the quarter. Subsequently, most of the students who had completed the dormitory and classroom questionnaires consented to have the student health center release information regarding the number of visits they made

[1]This assumption is consistent with social learning theory, which postulates that one's general expectations concerning the quality of interaction with others will be determined largely by his or her interpersonal experiences in specific situations (cf. Duke & Nowicki, 1972; Rotter, 1966; Rotter, Chance, & Phares, 1972).

there during the academic year. They also reported the number of times they had consulted off-campus physicians during the year.

The residential and classroom questionnaires incorporated items pertaining to both physical and social features of the environment. On the basis of questionnaire responses an attempt was made to predict sensitivity to crowding in the classrooms as a function of subjects' ratings of their residential environment. Moreover, the classroom responses of students reporting neutral and personal crowding experiences in their residences were compared in order to detect possible differences in the generalizability of crowding sensitivity from residential to classroom settings. Grades achieved by the students were recorded to determine possible effects of crowding on classroom performance.

The present investigation differs from earlier studies of dormitory crowding in an important respect. Previous studies have focused on the behavioral effects of exposure to different levels of social density, or group size, within residential situations.[2] Thus, the recent research conducted by Aiello, Epstein, and Karlin (1975) and Baron, Mandel, Adams, and Griffen (1975) examines the psychological, social, and health consequences of residing in tripled-up rooms; that is, where three students occupy a dormitory room originally designed for two. And the studies reported by Baum and Valins (1973) and Valins and Baum (1973) focus on the differential effects of corridor-design (34 residents per corridor) versus suite-design (6 residents per suite) dorms on the social behavior of their occupants.[3]

In contrast, the current study focuses on the psychological processes that may mediate one's sensitivity to crowding and the behavioral concomitants of this experience, irrespective of existing levels of physical or social density in the residential setting. Thus, one of the major concerns of this research is whether or not the subjective experience of crowding in the context of negative feelings about one's roommates is more highly predictive of crowding sensivivity in nonresidential situations, medical complaints, and impaired academic performance than is perceived crowding that is not accompanied by negative feelings toward others. We are interested in knowing, for example, not only whether the residents of corridor- and suite-design dorms differ in their perceptions of crowding, but also whether the residents within each of these groups are more adversely affected by this experience when they are not compatible with their roommates.

---

[2]Physical density refers to the amount of space available to a particular number of people, whereas social density denotes the number of people occupying a fixed amount of space.
[3]For more extensive reviews of this research, see Baum and Valins (1977), Stokols (1978), and Sundstrom (1978).

An examination of the behavioral and health-related concomitants of perceived crowding, rather than of social or physical density *per se*, seems important for several reasons. First, a number of experimental studies have shown that perceptions of crowding can vary independently of density levels as a function of certain nonspatial factors, e.g., personality characteristics, the quality of social relations existing among the occupants of an area, and situation-specific needs for privacy or solitude (cf. Altman, 1975; Stokols, 1976). Moreover, a comprehensive survey study of Toronto families conducted by Booth (1975) suggests that any behavioral and health-related outcomes that occur in high-density residential environments are more strongly associated with the subjective experience of crowding than with the levels of density existing within (persons per room) and outside (persons per acre) the household. Thus, the perception of crowding, in certain instances at least, appears to correlate significantly with important indices of personal well-being; it therefore seems important to learn more about the social and psychological circumstances that affect the intensity and impact of this experience.

The specific predictions of the study were as follows. (1) In a stepwise multiple regression analysis the prediction of classroom crowding will be significantly more reliable when ratings of both the physical and social dimensions of the residential environment are employed as predictor variables than when assessments of the physical environment are used alone. (2) Subjects whose residential evaluations indicate a pattern of personal crowding will express greater feelings of crowding in the classroom than those whose dormitory or apartment ratings reflect the experience of neutral crowding. (3) Students' visits to health centers on and off campus during the academic year and their course grades during the fall quarter will be significantly associated with their ratings of residential crowding and their evaluations of the physical and social conditions within their residences, as reported by them during the fall quarter.

The Study

Subjects

Participants in the study were drawn from a large undergraduate course (approximately 400 students) offered at the University of California, Irvine, during the fall quarter. From a listing of students taking the course, prospective subjects were identified on the basis of two main considerations: (1) place of residence and (2) year in college. An attempt was made to compose the sample primarily of 1st-year students living in suite-design dormitories, so as to minimize subjects' prior exposure to classrooms on the Irvine campus and to control for variations in residential density and design. Participants were drawn from the same lecture class to control for the effects of professorial style, course content, and classroom design on the data.

From an original listing of 60 prospective subjects, 21 volunteered to participate in the study and completed both the residential and classroom questionnaires. In view of the small sample size, an additional group of second-year students, some of whom resided in off-campus apartments, was included in the sample. The sample employed in the statistical analyses of the residential and classroom data consisted of 20 females and 11 males.

Near the end of the spring quarter, 1975, an attempt was made to contact these same subjects in their dormitory suites or by telephone. Two of the subjects could not be traced and two others did not consent to the release of information regarding their visits to the student health center. This brought the sample size used in the prediction of doctors' visits to 27 subjects. Of these, 12 females and 6 males resided in double rooms of suite-type dormitories, 1 female occupied a single dormitory room, and 3 females and 5 males lived in off-campus residences. Three subjects had changed rooms within the dormitories since the fall quarter and 1 subject had changed his off-campus apartment.

Procedure

Prospective subjects were contacted by telephone during the second week of the quarter and informed about a "study of the reactions of college students to dormitory (or off-campus) living conditions at U.C. Irvine." They were told that if they agreed to participate in the survey they would fill out questionnaires concerning the physical and social attributes of their current residence. The caller further explained that the questionnaire session would last for approximately 20 minutes, and that participants would be paid $2.00 at the session. Students agreeing to participate in the study were asked to report to an office on campus during the second or third week of the quarter where they completed the residential questionnaire.

The collection of classroom data occurred during the fourth week of the quarter. The professor[4] distributed a questionnaire to all members of the class during the initial portion of a class period. The questionnaire

[4]The authors would like to express their appreciation to Ralph Catalano, who administered the classroom questionnaires to students enrolled in his Principles of Social Ecology course, and to Susan Miller, who assisted in coding the data.

was described as part of a research project being conducted by a faculty associate, the purpose of which was to learn more about students' reactions to different kinds of courses at the University of California, Irvine. It was emphasized that completion of the questionnaire was entirely voluntary, and that each student's responses on it would have nothing to do with the grade he or she received in the course. Students volunteering to assist in the project were asked to put their student identification number on all pages of the questionnaire to facilitate collation of the data. The professor also noted that questionnaire responses would remain anonymous and would be coded on computer cards for statistical use only.

The questionnaire contained several items regarding the relative advantages and disadvantages of large lecture classes in comparison with other types of courses—small seminars and large discussion classes, for example. Embedded among these items was a set of bipolar scales concerning the physical conditions and social climate of the present classroom. These were identical to the items included in the dormitory questionnaire.

When the students were approached again during the spring quarter, they answered several questions pertaining to their health and were asked to sign a consent form allowing the student health center to release information regarding the number of visits made by each student during the three quarters of the academic year.

Measures and Analyses

The residential questionnaire incorporated six sets of 7-point semantic differential scales. The first included eight items pertaining to subjects' perception of their dormitory suite or apartment in terms of the quality of its physical dimensions. Subjects were asked to indicate the degree to which their residence permitted privacy and was pleasant, comfortable, spacious, quiet, large, uncluttered, and cheerful. The second set of scales related to the perceived quality of social relationships existing among themselves and their suitemates (roommates). Subjects were asked to rate the degree of trust, competition, alienation, similarity, and hostility felt among themselves and other occupants of the residence, as well as the extent to which they tried to make each other feel secure, were considerate of each other's feelings, and confided in each other about personal problems. A third set of scales pertained to the degree of crowding and spatial restriction felt by subjects in their dormitory suite or apartment.

Three other sets of semantic differential scales were included in the questionnaire. Two contained items from Rotter's (1966) Internal-External Scale and Keniston's (1965) Short Alienation Scales, respectively. On the remaining set subjects were asked to rate "people in general" along 10 dimensions, including, for example, "harmful-beneficial," "hostile-friendly," "disturbing-calming," and "bad-good."

Two final assessments were incorporated into the residential questionnaire. First, subjects were asked to draw a map of their dormitory suite or apartment on a blank piece of paper. Second, they were requested to complete the Comfortable Interpersonal Distance Scale (CID), a paper-and-pencil measure of personal space needs (Duke & Nowicki, 1972).

The classroom questionnaire incorporated three sets of 7-point scales. On the first set of scales, subjects were asked to rate how crowded, restricted, threatened, insecure, and tense they "usually feel in the present classroom." The second and third item-clusters pertained to subjects' evaluation of the classroom in terms of its physical conditions and social atmosphere, respectively. These items tapped the same dimensions of environmental amenity as those reflected in comparable scales of the dormitory questionnaire. Finally, a number of open-ended filler items were included in the classroom questionnaire that required subjects to list the relative advantages and disadvantages of large versus small and medium-sized classrooms.

The spring questionnaire consisted of questions pertaining to the number of times the students had visited medical doctors on or off campus during the academic year. The student health center located on the University of California campus reported the actual number of visits made by each student during the three quarter terms.

Three major analyses were performed on the data. First, the correlations between residential and classroom assessments of environmental quality were examined. The units of analysis were item-cluster total scores, which were computed by summing subjects' responses on the various scales within a particular cluster. Thus, each of the three cluster scores obtained from the residential data was correlated with each of the three total scores derived from the classroom data. The correlations between environmental evaluation scores and the additional indices included in the residential questionnaire also were examined.

Second, a series of stepwise regression analyses were performed on the dormitory, classroom, and health center data. In the primary analysis the classroom measure of felt crowding was employed as the response variable while cluster scores pertaining to the physical quality of the residential environment, its social atmosphere, and the perception of residential crowding were used as predictor variables. In a subsequent analysis assessments of subjects' internal-externality, chronic alienation, interpersonal distance preferences, and residential map size were incorporated into the regression equation as predictor variables.

In two other analyses course grades and visits to the student health center during the year were used as dependent variables, and residential crowding, physical conditions, and social atmosphere were employed as predictor variables in each. The classroom social atmosphere index was employed as an additional independent variable in the analysis of course grades.

Third, multivariate analyses of variance (MANOVA) were performed on the classroom data, utilizing the major indices of residential evaluation as blocking factors. In the first analysis subjects were divided into "low crowding" and "high crowding" groups through a median split of their residential crowding total scores. For the second analysis subjects' evaluations of the physical environment (EPE) and social environment (ESE) were utilized jointly to distinguish among four different groups of respondents: (1) those who evaluated both the physical and social dimensions of their residence positively (+EPE/ESE); (2) those who evaluated both dimensions negatively (−EPE/ −ESE); (3) those who reacted positively to the physical conditions of their dormitory suite or apartment but negatively to its social climate (+EPE/ −ESE); and (4) those who reacted negatively to the physical conditions and positively to the social climate of their residential environment (−EPE/+ESE). On the basis of these groupings, a two-factor (EPE × ESE) MANOVA was performed on the three major classroom total scores.

The EPE/ESE taxonomy of subjects provided a means of identifying individuals whose feelings of crowding were correlated with negative reactions to only the physical attributes of their residence, and those whose perceptions of crowding were associated with negative reactions to both the physical and social conditions within their residence. The differentiation between these groups was accomplished through a median split of subjects' crowding total scores and the subsequent grouping of these scores on the basis of subjects' EPE/ESE response patterns. It should be noted that the −EPE/+ESE and −EPE/ −ESE configurations of data correspond to patterns of neutral and personal crowding, respectively. Thus, by comparing the classroom crowding scores of these two groups through MANOVA procedures, it was possible to assess the relative degree to which neutral and personal crowding experiences in the residential environment affected subjects' sensitivity to crowding in the classroom.

Findings

Correlational Analyses

Intercorrelations among the major predictor and criterion variables are presented in Table 1. It is evident that assessments of the residential

Table 1.   Correlations among Major Predictor and Criterion Variables[a]

Variable	Residential crowd	EPE	ESE	I-E	Alienation
Residential crowding					
EPE	$-.82^d$				
ESE	$-.39^b$	$.53^d$			
I-E	$-.08$	$.17$	$-.03$		
Alienation	$.23$	$-.24$	$-.36^b$	$.28$	
People	$-.42^c$	$.32^b$	$.50^c$	$-.25$	$-.42^c$
CID	$-.07$	$.07$	$.09$	$.20$	$.04$
Map size	$-.03$	$.12$	$.35^b$	$-.28$	$-.34^b$
Classroom crowding	$.59^d$	$-.45^c$	$-.03$	$-.07$	$-.11$
Classroom physical	$-.45^c$	$.35^b$	$.07$	$.08$	$.11$
Classroom social	$-.47^c$	$.33^b$	$-.12$	$.05$	$.32^b$
Classroom security	$-.57^d$	$.50^c$	$.19$	$-.04$	$-.02$
Course grade	$.15$	$-.07$	$-.15$	$.14$	$.36^b$

[a]Pearson correlation coefficients provided.
[b]$p < .05$.
[c]$p < .01$.
[d]$p < .001$.

physical environment were highly correlated with those of physical and
social conditions in the classroom. Residential crowding, for example,
was significantly related to classroom crowding ($r_{31} = .59$, $p < .001$),
physical amenity ($r_{31} = -.45$, $p < .01$), social atmosphere ($r_{31} = -.47$, $p
< .01$), and security ($r_{31} = -.57$), $p < .001$). The total score for evaluation
of the residential physical environment reflected a similar pattern of
correlations with the classroom variables. Additionally, size of subjects'
residential map (measured by the area covered on an 8 1/2 × 11 inch
sheet of paper) was significantly correlated with classroom crowding
($r_{31} = .30$, $p < .05$), social atmosphere ($r_{31} = -.45$, $p < .01$), and security
($r_{31} = -.41$, $p < .01$).

While the index of residential social atmosphere was not correlated
with classroom total scores, it was significantly related to residential
crowding ($r_{31} = -.39$, $p < .05$) and physical amenity ($r_{31} = .53$,
$p < .001$). Also, subjects' ratings of people in general were significantly
correlated with residential crowding ($r_{31} = -4.2$, $p < .01$), physical
amenity ($r_{31} = .32$, $p < .05$), and social atmosphere ($r_{31} = .50$, $p < .01$).
Although alienation from people in general was not correlated with
either residential or classroom crowding, an index of perceived alienation from roommates (which was included in the ESE total score) was
significantly associated with residential crowding ($r_{31} = .57$, $p < .001$).
Finally, the summary index of classroom social climate was highly corre-

People	CID	Map size	Classroom crowding	Classroom physical	Classroom social	Classroom security
−.27						
.23	.17					
.13	−.05	.30^b				
−.18	.08	−.17	−.39^b			
−.28	−.08	−.45^c	−.25	.50^c		
.16	−.19	−.41^c	−.33^b	.33^b	.56^d	
−.37^b	.02	−.19	−.07	.18	.35^b	.40^b

lated with assessments of classroom physical conditions ($r_{31} = .50$, $p < .01$) and security ($r_{31} = .56$, $p < .001$).

Regression Analyses

In the initial regression analysis the total scores for residential crowding, physical conditions, and social atmosphere were utilized to predict classroom crowding. All three indices contributed significantly to prediction of the criterion variable ($F_{3,27} = 5.81$, $p < .01$). The analysis accounted for a total of 39% of the variance (see Table 2).

In the second analysis additional predictor variables were incorporated, namely, residential map size, I-E score, CID score, and alienation from others in general. Again, the indices of residential crowding, phys-

Table 2. Summary Table of Stepwise Multiple Regression Prediction of Classroom Crowding by Indices of Residential Evaluation[a,b]

Residential index	Step	Multiple r	Cumulative r^2	Simple r	Beta
Perceived crowding	1	.588	.346	.588	.635
Evaluation of the social environment	2	.625	.391	−.031	.246
Evaluation of the physical environment	3	.626	.392	−.450	−.060

[a]Simple correlation of residential index with criterion variable.
[b]Reliability of regression: $F = 5.81$; $df = 3,27$; $p < .01$.

Table 3. Summary Table of Stepwise Multiple Regression: Prediction of Classroom Crowding by Indices of Residential Evaluation and I–E, Alienation, CID, and Map Scores[a,b]

Index	Step	Multiple r	Cumulative r^2	Simple r	Beta
Perceived crowding	1	.588	.346	.588	.619
Evaluation of the social environment	2	.626	.391	−.031	.180
CID score	3	.634	.402	−.131	−.147
Map size	4	.645	.416	.149	.142
Evaluation of the physical environment	5	.646	.417	−.450	−.078
I–E score	6	.647	.419	−.097	.046

[a]Criterion variable: perception of crowding in the classroom.
[b]Reliability of regression: $F = 2.88$; $df = 6,24$; $p < .05$.

ical conditions, and social atmosphere contributed significantly to the prediction of classroom crowding, as did map size, I–E, and CID scores ($F_{6,24} = 2.88$, $p < .05$). A total of 42% of the variance was accounted for in the second analysis (see Table 3). It is evident from Tables 2 and 3 that residential crowding alone accounted for most of the variance (35%) in both analyses.

In the third regression analysis the classroom social atmosphere measure predicted the grades achieved in the course, with residential crowding and the subjects' evaluation of both the physical and social dimensions of their residences improving the prediction significantly ($F_{4,26} = 3.44$, $p < .05$) (see Table 4). And in the final regression analyses all three residential indices, crowding, physical conditions, and social atmosphere, contributed significantly to the prediction of visits to the student health center ($F_{3,23} = 8.07$, $p < .005$) and total visits to health centers on and off campus during the academic year ($F_{3,23} = 5.53$, $p < .01$)[5] (see Tables 5 and 6).

[5]An additional regression analysis was performed on the health center data using only those subjects who occupied architecturally comparable dorm rooms (10 feet 9 inches × 17 feet 1 inch) and had remained in the same suite throughout the entire academic year ($N = 15$). This analysis was intended to control for the effects of architectural factors on the relationship between perceived quality of the residential environment and total visits to the student health center during the year. As in the previous analyses, perceptions of residential crowding, physical conditions, and social climate were significantly related to total student health center visits during the academic year ($F_{3,11} = 3.96$, $p < .05$).

Table 4. Summary Table of Stepwise Multiple Regression: Prediction of Course
Grade by Indices of Residential and Classroom Evaluation[a,b]

Index	Step	Multiple r	Cumulative r^2	Simple r	Beta
Classroom social climate	1	.290	.084	.290	.415
Perceived residential crowding	2	.435	.190	.151	.975
Evaluation of the residential physical environment	3	.579	.336	.067	.756
Evaluation of the residential social environment	4	.589	.346	−.148	−.101

[a]Criterion variable: course grade. (Other variables not entered into the equation were map size, CID
score, classroom crowding, classroom physical environment, and I-E).
[b]Reliability of regression: $F = 3.44$; $df = 4,26$; $p < .05$.

Table 5. Summary Table of Stepwise Multiple Regression: Prediction of Visits
to Student Health Center during Academic Year by Indices of
Residential Evaluation[a,b]

Residential index	Step	Multiple r	Cumulative r^2	Simple r	Beta
Perceived crowding	1	.496	.246	.496	.781
Evaluation of the social environment	2	.615	.378	.095	.496
Evaluation of the physical environment	3	.716	.513	−.493	−.534

[a]Criterion variable: total visits to student health center during academic year.
[b]Reliability of regression: $F = 8.07$; $df = 3,23$; $p < .005$.

Table 6. Summary Table of Stepwise Multiple Regression: Prediction of Total
Visits to Health Centers, On and Off Campus, during Academic Year
by Indices of Residential Evaluation[a,b]

Residential index	Step	Multiple r	Cumulative r^2	Simple r	Beta
Perceived crowding	1	.527	.278	.527	.290
Evaluation of the social environment	2	.580	.337	−.025	.441
Evaluation of the physical environment	3	.647	.419	−.508	−.545

[a]Criterion variable: total visits to health center, on and off campus, during academic year.
[b]Reliability of regression: $F = 5.53$; $df = 3,23$; $p < .01$.

Multivariate Analyses of Variance

In the first analysis a one-way (residential crowding) MANOVA was performed on the classroom total scores of high- and low-crowding subjects. Results indicated that subjects who felt crowded in their residence rated the classroom environment more negatively than did low-crowding subjects (multivariate $F_{4,26} = 3.97$, $p < .01$). Univariate analyses revealed that high-crowding subjects felt more crowded ($F_{1,29} = 9.53$, $p < .004$) and less secure ($F_{1,29} = 13.53$, $p < .001$) in the classroom than did low-crowding subjects. The former group also rated the physical conditions ($F_{1,29} = 4.63$, $p < .04$) and social climate ($F_{1,29} = 4.52$, $p < .04$) of the classroom more negatively than did the latter. An additional MANOVA indicated that high-crowding subjects rated people in general more negatively than did low-crowding subjects ($F_{1,29} = 8.34$, $p < .007$).

A two-way (EPE × ESE) MANOVA on the classroom total scores revealed that low-EPE subjects (those who rated the residential physical environment negatively) felt more negative about the classroom environment than did high-EPE subjects (multivariate $F_{4,24} = 4.23$, $p < .008$). Univariate analyses indicated that low-EPE subjects felt more crowded ($F_{1,27} = 11.38$, $p < .002$) and less secure ($F_{1,27} = 14.16$, $p < .001$) than did high-EPE subjects, and also evaluated the physical features ($F_{1,27} = 4.21$, $p < .05$) and social atmosphere ($F_{1,27} = 5.99$, $p < .021$) of the classroom more negatively. Neither a significant main effect for ESE nor an EPE × ESE interaction effect was obtained.

In a final analysis, the classroom crowding scores of −EPE/+ESE and −EPE/−ESE subjects within the high residential crowding group were compared to detect possible differences in the generalization of neutral and personal crowding experiences from one situation to another. Results indicated that the means of the two groups were not significantly different.

Conclusions

The results of this research suggest that crowding experiences in residential settings are highly predictive of sensitivity to crowding in at least certain nonresidential environments. The correlation and regression analyses, as well as the MANOVAs, provide strong evidence that perceived crowding at home and negative feelings about the residential physical environment are associated with unfavorable reactions to both the physical and social dimensions of nonresidential settings.

In support of our first prediction the regression data revealed that subjects' evaluation of the social atmosphere of their residence contributed significantly to the prediction of their cross-situational sensitivity to crowding. This finding provides some support for a basic assumption underlying the proposed typology of crowding experiences, namely, that social factors as well as spatial variables mediate the perception of crowding and the generalization of crowding experiences from one situation to another. Similarly, Baum, Harpin, & Valins (1975) found that residents of corridor-design dorms who rated their social environment as "cohesive" reported lower levels of residential crowding than did those who rated their dorms as noncohesive.

The association between social variables and felt crowding also was reflected in the significant correlations between residential crowding and ratings of people in general, alienation from roommates, and perceived social climate of the residence. In line with these data, a recent experiment by Stokols and Resnick (1975) indicated that heightened levels of perceived evaluation by others and threat to personal security were associated with increased interpersonal distance and elevated ratings of subjective crowding in a laboratory situation.

The second prediction, that subjects in the personal-crowding residential group would express greater feelings of crowding in the classroom than those in the neutral-crowding group, was not supported by the MANOVA data. The absence of significant differences in classroom crowding between these groups may indicate that personal crowding experiences are not more generalizable between settings than are neutral crowding experiences; or, alternatively, the absence of significant between-groups differences may be related to certain features of the present study. First, small sample size limited the representativeness of our EPE/ESE groupings and the reliability of the related statistical analyses. Second, the fact that ratings of the residential social environment were obtained during the first few weeks of the fall quarter may have limited the extent to which social features of the residence could exert a significant impact on subjects' sensitivity to crowding in nonresidential settings. In view of the above limitations of this study, the prediction of differential generalization among neutral- and personal-crowding experiences remains to be more adequately examined through additional investigations.

The third prediction, that students' visits to physicians throughout the academic year and the quality of their course work during the fall quarter would be significantly associated with ratings of residential crowding, social climate, and physical conditions, was supported by the data. Students' visits to health centers on and off campus as well as their course grades obtained during the fall quarter were significantly predicted by subjective ratings of the residential environment, although the quality of course performance was more highly correlated with a measure of social atmosphere in the classroom than with the residential measures.

While the obtained pattern of health and academic performance data generally supports our third prediction, it does not provide a basis for inferring causal connections among residential crowding, classroom performance, and student medical complaints. Certainly, it is possible that residential crowding experiences promote medical problems and poor academic performance, but it is equally plausible that chronically unhealthy or unsuccessful students are more susceptible to crowding experiences and social problems than their healthy or academically gifted counterparts. At the same time, the significant inverse correlation observed between perceived residential crowding and positive feelings toward people in general ($r = -.42$, $df = 30$, $p < .01$) can be interpreted as evidence either that exposure to residential crowding promotes a negative view of other people or, alternatively, that individuals who are inherently antisocial are more likely to feel crowded in most situations than those who tend to react favorably toward others (see Table 1). To provide a straightforward test of our hypotheses, then, a longitudinal study is required in which students who are similar in terms of their prior health, interpersonal orientation, and scholastic achievement patterns can be interviewed at different times during the year to determine whether their academic and medical records are systematically affected by exposure to crowded living conditions. It also would be important to match the students with regard to the architectural features of their residences (e.g., amount and arrangement of space within dormitory suites). A follow-up study that incorporates the above-mentioned design features is currently in progress at the University of California, Irvine.

The results of this study suggest some additional directions for future research. First, the fact that perceived crowding was more highly associated with social variables within the residential environment than those within the classroom setting suggests that the contribution of social factors to crowding experiences may increase as a function of the "primariness" of the environment. Because subjects spend more time performing personally important activities in their residences than in a given classroom, the impact of socially mediated interferences on feelings of crowding would probably be greater in the former setting than in the latter. In order to explore the differential association between feelings of crowding and social dimensions of the environment, it will be necessary to develop a means of locating various environments along the primary-secondary continuum, and to sample the distribution of personal versus neutral crowding experiences within a diversity of settings.

In developing criteria for the coding of diverse environments it is important to recognize that the primary-secondary distinction implies both functional and experiential dimensions. Although most persons spend more time engaging in personally important activities within residential vis-à-vis nonresidential settings, individual judgments as to the importance of one's activities and the degree of personal investment in a particular setting are bound to vary. Thus, subjective-report as well as observational criteria should be employed in attempting to distinguish among primary and secondary environments.

An important task for future research will be to determine more precisely the conditions under which crowding experiences result in both immediate and cumulative behavioral impairments.

Acknowledgments

The authors would like to thank Jack Aiello, Andy Baum, Hans Esser, Gary Evans, and Gilbert Geis for their comments on an earlier version of the manuscript. Also, we appreciate the assistance of Gerald B. Sinykin, director of the Student Health Service at the University of California, Irvine, during the data-collection phase of this research.

References

Aiello, J., Epstein, Y., & Karlin, R. *Field experimental research on human crowding*. Paper presented at the Eastern Psychological Association Convention, New York, 1975.

Altman, I. *The environment and social behavior: Privacy, personal space, territory and crowding*. Monterey, California: Brooks/Cole, 1975.

Baron, R., Mandel, D., Adams, C., & Griffen, L. *Effects of social density in university residential environments*. Paper presented at the Annual Convention of the American Psychological Association, Chicago, 1975.

Baum, A., & Valins, S. Residential environments, group size, and crowding. *Proceedings of the 81st Annual Convention of the American Psychological Association*, 1973, 211–212.

Baum, A., & Valins, S. *Architecture and social behavior: Psychological studies of social density*. Hillsdale, New Jersey: Erlbaum, 1977.

Baum, A., Harpin, R. E., & Valins. S. The role of group phenomena in the experience of crowding. *Environment and Behavior*, 1975, 7, 185–198.

Booth, A. *Final report: Urban crowding project*. Paper presented to the Ministry of State for Urban Affairs, Government of Canada, 1975.

Desor, J. Toward a psychological theory of crowding. *Journal of Personality and Social Psychology*, 1972, 21, 79–83.

Dooley, B. B. Effects of social density on men with "close" or "far" personal space. *Journal of Population*, 1978, 1, 251–265.

Duke, M., & Nowicki, S. A new measure and social-learning model for interpersonal distance. *Journal of Experimental Research in Personality*, 1972, *6*, 119–132.

Esser, A. A biosocial perspective on crowding. In J. Wohlwill & D. Carson (Eds.), *Environment and the social sciences: Perspectives and applications*. Washington, D.C.: American Psychological Association, 1972.

Evans, G. *Physiological and behavioral consequences of crowding*. Unpublished doctoral dissertation, University of Massachusetts, Amherst, 1975.

Hanson, L., & Wicker, A. *Effects of overmanning on group experience and task performance*. Paper presented at Western Psychological Association Convention, Anaheim, California, April 1973.

Keniston, K. *Short alienation scales*. Unpublished manuscript, Yale University Medical School, New Haven, Connecticut, 1965.

Milgram, S. The experience of living in cities. *Science*, 1970, *167*, 1461–1468.

Proshansky, H., Ittelson, W., & Rivlin, L. Freedom of choice and behavior in a physical setting. In H. Proshansky, W. Ittelson, & L. Rivlin (Eds.), *Environmental psychology: Man and his physical setting*. New York: Holt, Rinehart & Winston, 1970.

Rodin, J. Crowding, perceived choice, and response to controllable and uncontrollable outcomes. *Journal of Experimental Social Psychology*, 1976, *12*, 564–578.

Rotter, J. Generalized expectancies for internal vs. external control of reinforcement. *Psychology Monographs*, 1966, *80*, (Whole No. 609).

Rotter, J., Chance, J., & Phares, E. *Applications of a social learning theory of personality*. New York: Holt, Rinehart & Winston, 1972.

Sherrod, D. Crowding, perceived control and behavioral after-effects. *Journal of Applied Social Psychology*, 1974, *4*, 171–186.

Stokols, D. On the distinction between density and crowding: Some implications for future research. *Psychological Review*, 1972, *79*, 275–277. (a)

Stokols, D. A social psychological model of human crowding phenomena. *Journal of the American Institute of Planners*, 1972, *38*, 72–84. (b)

Stokols, D. Toward a psychological theory of alienation. *Psychological Review*, 1975, *82*, 26–44.

Stokols, D. The experience of crowding in primary and secondary environments. *Environment and Behavior*, 1976, *8*, 49–86.

Stokols, D. A typology of crowding experiences. In A. Baum & Y. Epstein (Eds.), *Human response to crowding*. Hillsdale, New Jersey: Erlbaum, 1978.

Stokols. D., & Resnick, S. *An experimental assessment of neutral and personal crowding experiences*. Paper presented at the Annual Conference of the Southeastern Psychological Association, Atlanta, March 1975.

Sundstrom, E. Crowding as a sequential process: Review of research on the effects of population density on humans. In A. Baum & Y. Epstein (Eds.), *Human response to crowding*. Hillsdale, New Jersey: Erlbaum, 1978.

Valins, S., & Baum, A. Residential group size, social interaction, and crowding. *Environment and Behavior*, 1973, *5*, 421–439.

Wicker, A. Undermanning theory and research: Implications for the study of psychological and behavioral effects of excess populations. *Representative Research in Social Psychology*, 1973, *4*, 185–206.

Zlutnick, S., & Altman, I. Crowding and human behavior. In J. Wohlwill & D. Carson (Eds.), *Environment and the social sciences: Perspectives and applications*. Washington, D.C.: American Psychological Association, 1972.

8

Environmental Satisfaction in High- and Low-Rise Residential Settings: A Lewinian Perspective

Charles J. Holahan and Brian L. Wilcox

Lewin's classic formula, $B = f(P,E)$, has for almost 40 years remained a keystone in the framework of social psychological theory. The formula serves less as a precise mathematical function than as a working definition of a broad sphere of research and inquiry. Yet, despite its generality, it affords a conceptual backdrop for a broad spectrum of contemporary research concerned with investigating personal versus situational effects in predicting social behavior. The results of this research have consistently demonstrated that both personal and situational variables and, especially, the interaction between personal and environmental factors are of importance in accounting for the total behavioral variance (cf. Bem & Allen, 1974; Ekehammer, 1974; Endler & Hunt, 1968; Mischel, 1973; Raush, Dittmann, & Taylor, 1959; Raush, Farbman, & Llewellyn, 1960). The present study was concerned with applying such an interactional framework in investigating environmental satisfaction in high- and low-rise student housing.

Our recent work in a university dormitory setting provided a unique opportunity for investigating the "person versus situation" issue. Student housing on the local campus presented a compelling environment for psychological investigation since the campus boasted one of

Charles J. Holahan and Brian L. Wilcox • Department of Psychology, University of Texas at Austin, Austin, Texas 78712.

the largest student resident halls in the country—a megadorm of two towers, housing a total of more than 3,000 students. In addition, the megadorm setting had from the start been plagued by serious problems, involving student complaints, excessive property damage, and an increasing vacancy rate. Although our interest in student housing was initiated by a request for a psychological assessment of student dissatisfaction in the megadorm, our investigative concerns broadened to understanding personal versus situational variables as predictors of living satisfaction at a more general level.

Earlier Research on Residential Environments

A number of authors have stressed the impact that the university residential living environment has on student development (Chickering, 1972; Feldman & Newcomb, 1969; Newcomb, 1962, 1966). It has been posited that the immediate living environment may have a significant impact on students in areas such as intellectual productivity, satisfaction with college life, emotional development, and the development of interpersonal relationship skills. Given this, and the fact that for many students the university residence environment represents the dominant locale for their time, energy, and activity, Heilweil (1973) has stressed the need for evaluative research on the psychological impact of the university residential living environment. He adds that the alternative is to relinquish this responsibility to chance, intuition, and custodial rather than functional and psychological concerns.

A number of research studies concerned with quality of life in student residential environments have reported less living satisfaction and social cohesion in high-rise megadorms in contrast to low-rise dormitory settings. Findings have pointed to less prosocial behavior and cooperation (Bickman, Teger, Gabriele, McLaughlin, Berger, & Sunaday, 1973) and a lower level of students' perception of social support and cohesiveness (Wilcox & Holahan, 1976) in high-rise as opposed to low-rise student housing. Crowding in dormitory settings has been demonstrated to be related to increased stress along with decreased social contact (Valins & Baum, 1973), more negative ratings of living space (Eoyang, 1974), and more negative interpersonal attitudes (Baron, Mandel, Adams, & Griffen, 1976). While these findings from previous research are of considerable importance to university educators and administrators, a further question needs to be asked, which is of both practical and theoretical significance to the environmental psychologist. Are different types of students differentially affected by contrasting residential settings?

A Field-Based Evaluation

The purpose of the present study was to extend the investigation of residential satisfaction in university housing environments to include an interactional analysis, taking account of differences in both characteristics of students and types of residential settings. Specifically, the study involved (1) a comparison of residential satisfaction and friendship formation in high- and low-rise dormitory settings and (2) an analysis of the interaction between students' levels of social competence and type of environment in affecting both residential satisfaction and friendship formation. It was predicted that (a) according to previous research in student residential environments, high-rise dormitories would be characterized by lower levels of satisfaction and friendship development than would low-rise dormitories; and (b) based on earlier interactional studies, the interactions between type of environment and social competence would be of particular importance in predicting both satisfaction and friendship formation.

Subjects in the study were 129 randomly selected second-semester freshmen who were residents of on-campus university housing at a state university. The sample included 55 males and 74 females. Only freshmen were used in order to limit self-selection for specific living arrangements likely to occur for advanced students with prior living experience on campus. Freshmen on the present campus were typically randomly assigned to the different dormitory settings, and where such random assignment clearly did not exist, residents were excluded from the present analysis. Nevertheless, the sample cannot be assumed to be entirely random, and in order to examine for any possible systematic population bias across dormitories, the subject samples from each dormitory were compared along a number of demographic indicators (family economic level, racial distribution, geographic background) and the social competence measure. There were no significant differences between dormitories on any of these measures.

Two types of student residential environments were compared in the study, a megadorm characterized by a high level of residential density and a number of low-rise dormitories reflecting a low level of residential density. The megadorm consisted of a tower of 10 floors occupied by males and a tower of 13 floors occupied by females. The megadorm housed approximately 3,000 students. The low-rise dormitories consisted of two dormitories for males (one of two floors and one of four floors) and two dormitories for females (one of two floors and one of five floors). Each low-rise dormitory housed approximately 250 students.

Evaluation Measures

Social Competence. Since the interactionist perspective stresses that individual characteristics interact with environmental factors in predicting living satisfaction, we decided to measure level of social competence under the assumption that this personal trait would be especially important in coping with the demands of residential life. Social competence was measured by the Texas Social Behavior Inventory (TSBI) developed by Helmreich, Stapp, and Ervin (1974). The authors report that the TSBI offers a highly reliable instrument for assessing individuals' stable self-perception of social competence and social self-esteem. They note that the instrument demonstrates a high level of construct validity based on its relationship with other measures of ability, attitude, and personality. For example, the TSBI has been found to correlate highly ($r = .50$, $p < .001$) with the self-esteem scale of the California Personality Inventory (Gough, 1964). Factor analysis has yielded factors of confidence and dominance along with social competence. Representative items from the 16-item (5-point scale) TSBI are: I feel I can confidently approach and deal with anyone I meet; when I work on a committee I like to take charge of things; I feel confident of my appearance; I have no doubts about my social competence.

Sociometric Scale. Through discussions with dormitory staff, it was felt that for university freshmen new to the campus setting, the establishment of proximal and easily accessible friendship networks within the dormitory setting represented a particularly important component of positive adjustment to residential life. This assumption was reinforced through input from resident assistants that a chief complaint of dissatisfied residents in the megadorm concerned social isolation. Based on this information, we decided to develop a sociometric technique to measure the level of friendship networks established within the residential setting. We felt a valid measure should tap meaningful relationships in addition to casual social contacts. Thus, a measure of dormitory-based friendships was developed, oriented toward measuring friendships at three levels of intimacy: casual-recreational, personal-conversational, and supportive. The items that tapped each of these levels respectively asked the respondent to identify (1) a friend to join you in a casual outing to a film or a sports event, (2) a friend to join you in a personal conversation where attitudes and values are shared, (3) a friend to join you in discussing an intimate personal problem concerning your feelings about a member of your family. In responding to the scale, students were asked to choose up to five friends at each intimacy level who resided within their own dormitory. For quantification purposes, choices were weighted (from 5 for a first choice to 1 for a fifth choice) and a final score was obtained by summing over the three friendship levels.

Environmental Satisfaction. A measure of satisfaction with the living environment was developed, which was composed of 10 scaled (7-point) items. The items measured satisfaction with meeting people and making friends, recreation, opportunities and places for personal conversation, finding help or support for a personal problem, comfort, privacy, student influence in policy decisions, physical layout of building and rooms, furnishings, and overall feelings about living in the particular dormitory. The predictive validity of the satisfaction scale was determined by correlating the satisfaction score with students' decisions to remain in or leave the particular dormitory for the next academic year. A statistically significant correlation was obtained ($r = .41$, $p < .01$).

Procedure

The measures were administered in a survey which required approximately 15 minutes for completion. The surveys were handed out to each subject personally by the resident assistant on his or her floor. The subject was asked to complete the survey anonymously and to return it to the assistant's mailbox within two days. The survey was conducted simultaneously across all settings one month before the end of the spring semester. An existing financial policy at the university required all students to honor their residential lease for the full academic year, assuring that the residential drop-out rate was very low. The few students who did drop-out during the year tended to be academic drop-outs, and these were evenly distributed over type of residential setting. Trained experimental assistants working with the resident assistants assured a standardized administration procedure. One hundred and twenty forms were administered in the male and female low-rise dorms respectively. The return rate was 62% in the megadorms and 58% in the low-rise dorms. (In order to decrease disproportionality in cell size, 32 subjects were removed in a random manner from three cells with disproportionately high n's.)

Results

Table 1 shows mean responses by dormitory, sex, and social competence on the satisfaction and friendship measures. In the statistical analysis social competence was treated as a blocking variable with high and low groups established through a median split. Table 2 summarizes

Table 1. Mean Scores by Dormitory, Sex, and Competence on the Residential Satisfaction and Friendship Formation Measures

	Males				Females			
	High competent		Low competent		High competent		Low competent	
Measures	X	SD	X	SD	X	SD	X	SD
Low-rise dormitories								
Residential satisfaction	54.23	9.92	42.56	8.16	51.75	14.62	48.35	12.32
Friendship formation	15.00	8.61	12.11	7.64	20.90	8.85	21.00	7.65
Megadorm								
Residential satisfaction	39.34	10.54	38.00	8.00	36.79	13.33	45.75	10.76
Friendship formation	16.32	10.69	17.36	7.89	11.14	8.61	21.35	9.07

Table 2. Results of Three Factor Multivariate Analysis of Variance (Dormitory × Sex × Competence) with Residential Satisfaction and Friendship Formation as Dependent Variables

Source	Multivariate F ($df = 2/120$)	Univariate F's ($df = 1/121$)	
		Satisfaction	Friendship formation
Dormitory (A)	9.71^b	19.47^b	1.46
Sex (B)	2.97^c	.55	5.92^a
Social competence (C)	1.28	.04	2.31
A × B	3.80^a	.04	7.10^b
A × C	3.43^a	4.92^a	3.37^c
B × C	4.14^a	5.95^a	4.05^a
A × B × C	.37	.36	.23

$^a p < .05.$
$^b p < .01.$
$^c p < .07.$

the results of a 2 × 2 × 2 multivariate analysis of variance (dormitory × sex × social competence) with residential satisfaction and friendship formation as dependent variables. For the multivariate F's, significant effects were found for the dormitory main effect and for all three two-way interactions.

For the univariate analyses, two significant main effects were found. Residents of the low-rise dormitories scored significantly higher on the measure of residential satisfaction than did residents of the megadorm. On the friendship measure, female residents scored significantly higher than did males. In addition, a number of interesting two-way interactions occurred in the univariate analyses.

On the satisfaction measure, two interactions were significant (dormitory × social competence, sex × social competence). In the low-rise dormitories, high competent students were more satisfied than low competent ones, while in the megadorm, in contrast, low competent individuals were slightly more satisfied than high competent ones. For male residents, social competence was positively related to satisfaction, while for females social competence showed an inverse relation to satisfaction.

On the friendship measure, two interactions were significant (dormitory × sex, sex × social competence), and another showed a statistical trend at the .07 level (dormitory × social competence). In the low-rise dormitories, female students scored markedly higher on the index of friendship formation than did males, while in the megadorm, in contrast, there was no relationship between sex and friendship formation.

For female residents, social competence showed a strong negative relationship to friendship formation, while for males, no relation was observed between social competence and dormitory-based friendships. Also, in the megadorm, low competent subjects scored higher on the friendship scale than did high competent ones, while in the low-rise dormitories the reverse was true.

Social Climate in High- and Low-Rise Settings

These results support the negative picture of residential life in high-rise as opposed to low-rise dormitory settings reported in earlier research (Baron *et al.*, 1976; Bickman *et al.*, 1973; Valins & Baum, 1973; Wilcox & Holahan, 1976). Residents of the megadorm were significantly more dissatisfied than residents of the low-rise dormitories. The range of dissatisfaction in the megadorm was very broad, including feelings about social contact and support, features of the physical environment, and student involvement in policy decisions.

These findings raise the question of how the social and organizational climates of the high-rise and low-rise residence halls may differ. To examine this question, we administered the University Resident Environment Scale or URES (Gerst & Moos, 1972) to a new sample of 110 freshmen (53 males and 57 females) in the two residential settings. The URES consists of 100 true–false items, which make up 10 subscales. These subscales are sorted into three underlying dimensions. The subscales of Involvement and Support compose the relationship dimensions, which deal generally with the modes of interpersonal relationships characteristic of the setting. The subscales of Independence, Traditional Social Orientation, Competition, Academic Achievement, and Intellectuality make up the personal growth or development dimensions. These dimensions assess the press toward personal and intellectual maturation. The final three subscales, Order and Organization, Student Influence, and Innovation, compose the system maintenance and system change dimensions, which deal with the impact of the formal organizational structure on student life.

Table 3 shows the mean score and standard deviation for each of the 10 URES subscales for males and females in the megadorm and the low-rise dorm. Using an analysis of variance format, sex and dorm type were compared in a 2 × 2 analysis of variance with each of the 10 URES subscale scores as dependent variables. Five of the 10 main effects for building type were significant, with residents of the high-rise buildings rating their environments lower on Involvement, Support, Order and Organization, and Student Influence and higher on the Independence

Table 3. Mean Scores on Ten URES Subscales for Males and Females in Megadorm and Low-Rise Environments

	Megadorm				Low-rise dorm			
	Male		Female		Male		Female	
	$\bar{X}$	SD	$\bar{X}$	SD	$\bar{X}$	SD	$\bar{X}$	SD
Involvement	4.21	2.29	4.10	2.36	6.68	2.79	6.42	2.65
Support	4.00	2.14	4.82	3.04	5.62	2.50	7.10	2.31
Independence	6.36	1.47	6.28	1.80	5.06	2.48	3.39	1.99
Traditional social orientation	4.64	1.79	5.78	2.76	4.31	2.27	7.36	1.17
Competition	3.35	1.98	3.50	1.95	3.18	2.28	3.07	1.89
Academic achievement	4.35	2.48	5.18	2.24	5.62	2.15	5.00	2.05
Intellectuality	4.00	2.05	4.28	2.20	4.37	2.12	4.34	2.29
Order and organization	4.32	2.03	5.85	2.41	6.56	2.22	7.39	2.15
Student influence	3.42	1.45	3.57	1.87	4.81	1.55	4.36	1.63
Innovation	3.78	1.91	4.87	1.95	5.12	2.12	3.73	1.94

subscale. Thus, the megadorm environment relative to the low-rise dormitories demonstrated markedly lower ratings on relationship and system maintenance and change dimensions. Consistent with earlier findings (Gerst & Moos, 1972), university women rated their environments higher on Support, Traditional Social Orientation, and Order and Organization and lower on Independence than did men. Additionally, a number of Dormitory × Sex interactions were observed, reflecting a dramatically strong effect of the megadorm environment on women residents. On the Independence subscale, while men in general rated their environments significantly higher, women in the high-rise dorm actually rated their environment higher on this scale than did the low-rise male residents. While women in general tended to rate their environments higher on the Traditional Social Orientation subscale, women in the low-rise building rated their environments much higher than did the high-rise female residents. Finally, males in the high-rise dorm and females in the low-rise dorms perceived their environments as significantly more innovative than did males in the low-rise dorms and females in the high-rise dorm.

Taken together with previous research findings concerned with the quality of social life in high- versus low-rise university housing, the results of the present study make untenable the "educational logic" underlying the movement toward megadorm environments throughout the country. It is unfortunate and disturbing that considerations such as spatial constraints and financial exigency are often the sole criteria used by university planners. That physical environmental factors exert such an impact on the social environments they encompass suggests that greater attention must be given to the design of physical settings, for in constructing a building one is also designing an important segment in students' social psychological experiences at college.

Implications of Interactional Findings

One of the most important implications of the present findings is to underscore the importance of viewing student adjustment from an interactional perspective. This concern is congruent with increasing research evidence that adequate prediction of social behavior necessitates the measurement of both personality and situational variables (cf. Bem & Allen, 1974; Ekehammer, 1974; Endler & Hunt, 1968; Kjerulff & Wiggins, 1976; Mischel, 1973; Moos, 1973). Clearly, no single type of educational environment will be ideally suited to the needs of all students. Adequate educational planning will necessitate a fuller appreciation of how different types of students are likely to respond to and attempt to cope with contrasting environmental settings.

While female residents established more dormitory-based friendships than did males, an especially interesting finding was the interaction between sex and type of dormitory in affecting relationships, reflecting differential styles of responding to environmental settings as a function of sex. While present knowledge in this area is limited, Holahan and Holahan (1976) present data that relate to the present findings. In examining verbal descriptions of university dormitory environments, they found that females focused particularly on the reduced personalization of the dormitory setting, while males emphasized the increased social contact possible in a communal living environment. In the present study the reduced personalization in the megadorm relative to the low-rise dormitories may have discouraged friendship formation on the part of females. Males, in contrast, may have been more inclined to develop friendships in the megadorm than in the low-rise dormitories due to the relatively greater level of social contact that they permitted.

The interaction between social competence and residential density in affecting residential satisfaction was complex and somewhat surprising. The nature of the relationship between residential satisfaction and social coping may clarify this finding. In the highly dissatisfying megadorm environment, more socially competent residents may have elected not to establish dormitory-based relationships; thus, indirectly facilitating the development of friendships among less socially competent students. These friendships in the megadorm may in turn have provided a foundation for the higher level of residential satisfaction among less competent students. There are two sources of data that lend some support to this interpretation. The social coping measure was significantly related to residential satisfaction ($r =.28$, $p <.01$), and additional anecdotal data indicated a strong tendency for high socially competent residents in the megadorm to form more friendships outside of the dormitory setting than did less competent students. Of course, caution needs to be exercised in generalizing these findings to students in other university settings. The subject sample we have examined here was select both in terms of the characteristics of students who initially chose dormitory living at this university and in terms of those students who responded to the survey. Finally, it is possible that the negative attitudes toward the megadorm setting may reflect, in part, a feeling on the part of residents that it is "fashionable" to complain about high-rise dormitory living. However, the specificity of the satisfaction scale items provides some assurance against this interpretation.

Applying the Evaluation Findings

These findings from the evaluation survey were used as feedback to relevant campus administrators for deciding on policy or design alternatives, and as a baseline against which to measure the impact of any changes based on the feedback. Feedback sessions were held with staff in the megadorm environment, including the dormitory director, assistant director, and student-life coordinators. The first stage of the feedback process involved weekly meetings between the principal investigators and the primary clients of the project—the director of the megadorm setting and the assistant director in charge of student activities. Through these collaborative planning sessions a general goal for applying the findings emerged, which concerned a commitment toward improving the quality of life for student residents in the megadorm setting. Specifically, it was felt this would involve enhancing the opportunity for social interaction in the environment, increasing privacy, and encouraging design solutions that were more flexible, thus accommodating a greater diversity of students' behavioral styles. During the second stage of the feedback process, in addition to the weekly meetings with the director and the assistant director, a number of meetings were also held with the six coordinators of student activities who worked under the assistant director.

All of these feedback sessions were conducted in a collaborative fashion. First, findings from the study were presented and discussed. This was followed by a brainstorming session oriented toward generating a range of solution alternatives. Finally, possible solutions were reconsidered in the light of administrative priorities, financial constraints, likely success, and the time period necessary for implementation. The solutions emerging from this process were then accepted as workable strategies for improving student satisfaction that could be implemented over a relatively short-term period. A second goal on the part of the principal investigators during this phase was to train the administrative staff in a number of "metaskills," relevant to understanding and dealing effectively with the feedback findings. These metaskills involved learning to articulate behavioral objectives for a residential setting, to evaluate environmental performance in behavioral terms, and to employ user-based data in making ongoing administrative and policy decisions.

The particular change strategies that were developed were of two types: (1) administrative policy revisions oriented toward facilitating students' adaptive coping efforts with the high-rise environment, and (2) decisions concerning strategic physical remodeling in areas where minimal financial input might generate optimal social psychological benefits. Policy revisions included resident assistant training programs based on the study's findings, an emphasis on increased social contact between students and residence hall staff, and increased social functions at the floor level to facilitate the establishment of friendship networks. An important physical design change that emerged from the feedback sessions involved the partial remodeling of the communal dining room in the megadorm. A focal area of concern on the part of residential staff was the dining area, which, while potentially an important social setting in the living environment, had evolved into a highly institutionalized and socially isolated setting. A planned environmental change was effected, which involved the construction of partitions in the previously open-space dining setting, in an effort to improve the opportunity for social contact, increase privacy, and diminish the overall institutional appearance of the setting. In order to evaluate the psychological impact of the new partitions, we administered a short survey in a pre–post format and also observed behavior in both partitioned and nonpartitioned sections of the cafeteria after the remodeling. Results of the evaluation indicated that the partitioning of the dining area was highly successful in achieving the established objectives. Students under the partition arrangement consistently expressed more satisfaction with the dining area and demonstrated higher levels of social behavior than did students under the nonpartition arrangement.

Broader Implications for a Lewinian Perspective

While these findings are specific to the particular setting and variables examined here, some general implications do emerge from this study that bear on the broader issue of person versus situation in behavioral prediction. First, the results support including both personal and environmental variables in predictive behavioral models as proposed in Lewin's original formula. Second, while both person and environment are critical in the overall model, their relative importance may vary with the particular characteristics of the situations studied. Third, the present data underscore the value of examining further the role of adaptive social behaviors as mediating links between the independent and dependent variables, especially where the outcome behavior investigated has a strong adaptational or adjustive component.

Finally, the implications of the interactional perspective discussed here bear particular relevance for psychologists concerned with the application of research knowledge to the resolution of societal problems. Recent criticism (Caplan & Nelson, 1973) has stressed the unfortunate social consequences inherent in psychologists' applying exclusively person-centered theoretical models to the investigation of complex and multiply determined social issues. There is clear practical utility in conceptual paradigms of social behavior able to accommodate multivariate complexity (cf. McGuire, 1973), adaptive emphasis in terms of effective coping strategies, and results applicable to a range of intervention targets. After 40 years, a second of Lewin's guiding dictums remains compelling: "There is nothing so practical as a good theory."

References

Baron, R. M., Mandel, D. R., Adams, C. A., & Griffen, L. M. Effects of social density in university residential environments. *Journal of Personality and Social Psychology*, 1976, *34*, 434–466.

Bem, D. J., & Allen, A. On predicting some of the people some of the time: The search for cross-situation consistencies in behavior. *Psychological Review*, 1974, *81*, 506–520.

Bickman, L., Teger, A., Gabriele, T., McLaughlin, D., Berger, M., & Sunaday, E. Dormitory density and helping behavior. *Environment and Behavior*, 1973, *5*, 465–490.

Caplan, N., & Nelson, S. D. On being useful: The nature and consequences of psychological research on social problems. *American Psychologist*, 1973, *28*, 199–211.

Chickering, A. W. *Education and identity*. San Francisco: Jossey-Bass, 1972.

Ekehammer, B. Interactionism in personality from a historical perspective. *Psychological Bulletin*, 1974, *81*, 1026–1048.

Endler, N. S., & Hunt, J. McV. S-R inventories of hostility and comparisons of the proportions of variance from persons, responses and situations for hostility and anxiousness. *Journal of Personality and Social Psychology*, 1968, *9*, 114–123.

Eoyang, C. K. Effects of group size and privacy in residential crowding. *Journal of Personality and Social Psychology*, 1974, *30*, 389–392.

Feldman, K., & Newcomb, T. M. *The impact of college on students* (Vols. 1 and 2). San Francisco: Jossey-Bass, 1969.

Gerst, M. S., & Moos, R. H. The social ecology of university student residences. *Journal of Educational Psychology*, 1972, *63*, 513–525.

Gough, H. G. *California personality inventory: Manual.* Palo Alto: Consulting Psychologists Press, 1964.

Heilweil, M. The influence of dormitory architecture on resident behavior. *Environment and Behavior*, 1973, *5*, 377–412.

Helmreich, R., Stapp, J., & Ervin, C. The Texas Social Behavior Inventory (TSBI): An objective measure of self-esteem or social competence. *Journal Supplement Abstract Service Catalog of Selected Documents in Psychology*, 1974.

Holahan, C. J., & Holahan, C. K. Sex-related differences in the schematization of the behavioral environment. *Personality and Social Psychological Bulletin*, 1977, *3*, 123–126.

Kjerulff, K., & Wiggins, N. H. Graduate student styles of coping with stressful situations. *Journal of Educational Psychology*, 1976, *68*, 247–254.

McGuire, W. J. The yin and yang of progress in social psychology: Seven koan. *Journal of Personality and Social Psychology*, 1973, *26*, 446–455.

Mischel, W. Toward a cognitive social learning reconceptualization of personality. *Psychological Review*, 1973, *80*, 252–283.

Moos, R. Conceptualization of human environments. *American Psychologist*, 1973, *28*, 652–655.

Newcomb, T. M. Student peer-group influence and intellectual outcomes of college experience. In R. Sutherland, W. Holtzman, E. Koile, & B. Smith (Eds.), *Personality factors on campus*. Austin, Texas: Hogg Foundation for Mental Health, 1962.

Newcomb, T. M. Research on student characteristics: Current approaches. In L. Dennis & J. Kauffman (Eds.), *The college and the students*. Washington, D.C.: American Council on Education, 1966.

Raush, H. L., Dittman, A. T., & Taylor, T. J. Person, setting and change in the social interaction. *Human Relations*, 1959, *12*, 361–379.

Raush, H. L., Farbman, I., & Llewellyn, L. G. Person, setting, and change in social interaction: II. A normal control study. *Human Relations*, 1960, *13*, 305–333.

Stern, G. G. *People in context*. New York: Wiley, 1970.

Valins, S., & Baum, A. Residential group size, social interaction, and crowding. *Environment and Behavior*, 1973, *5*, 421–439.

Wilcox, B. L., & Holahan, C. J. Social ecology of the megadorm in university student housing. *Journal of Educational Psychology*, 1976, *68*, 453–458.

9

Crowding and Personal Control: Social Density and the Development of Learned Helplessness*

Andrew Baum, John R. Aiello, and Lisa E. Calesnick

Although the experimental literature on crowding has expanded dramatically, much of this work has focused on one or another of a number of conditions, all plausibly caused by high density. Crowding has been operationalized in terms of the amount of space available (e.g., Freedman, 1975; Stokols, 1972); close physical proximity (e.g., Aiello, Epstein, & Karlin, 1975; Worchel & Teddlie, 1976); behavioral constraint and interference (e.g., Schopler & Stockdale, 1977; Sundstrom, 1975); and increasing levels of social stimulation, unwanted interaction, and overload (e.g., Baum & Valins, 1977; Desor, 1972; Saegert, 1978). More recently, research has suggested that people respond differently to social and spatial antecedents (Baum & Koman, 1976; Stokols, 1976) and that studies emphasizing one of these conditions may not generalize to situations in which other conditions are salient. More integrative descriptions

*This paper was originally published in the *Journal of Personality and Social Psychology*, Vol. 36, No. 9, pp. 1000–1011. Copyright 1978 by the American Psychological Association. Reprinted by permission.

Andrew Baum • Department of Medical Psychology, Uniformed Services University, School of Medicine, Bethesda, Maryland 20014. **John R. Aiello** • Department of Psychology, Rutgers University, New Brunswick, New Jersey 08903. **Lisa E. Calesnick** • Department of Psychology, Trinity College, Hartford, Connecticut 06106.

of crowding, focusing on its control-debilitating effects, have appeared (e.g., Altman, 1975; Baron & Rodin, 1978), and since both social and spatial conditions associated with high density may affect an individual's ability to regulate social experience, the use of control constructs may provide a broader perspective with which to view crowding. At the very least, such a perspective should suggest interesting and important directions for future research.

Crowding and Helplessness

One such direction is the study of the relationships among density, crowding, and motivational deficits characteristic of learned helplessness (cf. Seligman, 1975). If crowding involves density-related loss of control, and helplessness is conditioned by repeated or prolonged exposure to situations perceived as uncontrollable (i.e., responses and outcomes are independent), then long-term crowding should be associated with helplessness.

Rodin (1976) has conducted two studies that support the notion that high density and helplessness are linked. Controlling for socioeconomic status, she found that children living in high-density residences were less likely to assume control of available outcomes than were children living in less dense settings. Density of junior high school students' homes also affected performance on puzzle tasks. High-density subjects performed more poorly on solvable tasks than did subjects from lower density homes after initial exposure to unsolvable problems.

Social Density and Regulatory Control

Baum and Valins (1977) also considered the role of social control in high-density residential settings. When residents were grouped around shared lounge, bathroom, and hall spaces in relatively large numbers (32–40), interaction with known and unknown neighbors was frequent. As a result, residents' ability to predict and determine the nature, frequency, and duration of interpersonal contact was impaired. Compared with students grouped around shared living spaces in smaller numbers (6–20), these students felt that they exerted less control over social experience and that their dormitories were more crowded (despite equivalent dormitory densities and comparable backgrounds of the two groups).

Large residential group size also appeared to be associated with social withdrawal and reduced motivation to control outcomes. In one experiment by Baum and Valins (1977), subjects were told that they could choose the experimental condition in which they would participate. Residents of long-corridor dormitory settings (large residential groups) were less likely to respond to this choice by seeking additional information or attempting to choose than were residents of the short-corridor dormitory (small residential groups).

A second study by Baum and Valins assessed responses to a modified Prisoner's Dilemma game. By providing a withdrawal option as well as cooperative and competitive choices, this kind of game has been used to assess motivational deficits associated with helplessness (Kurlander, Miller, & Seligman, cited in Seligman, 1975).[1] Short-corridor residents chose equally cooperative and competitive strategies and selected very few withdrawal responses, regardless of the likelihood of interaction with the other player. When interaction was likely, long-corridor residents responded very competitively, but competitiveness decreased and withdrawal responding increased when interaction was not probable. Since long-corridor residents persistently avoided social contact, competitiveness may have reflected attempts to reduce involvement with the other player. When interaction was unlikely, this was not necessary, and motivation to influence the game seemed to decrease.

Glass and Singer (1972) and Seligman (1975) have considered the motivational costs of exposure to uncontrolled or unpredictable stress, and these considerations may be relevant in describing the behavior of these dormitory residents. Helplessness "training" may begin when long-corridor residents realize that their attempts to regulate social contact are ineffective and have little impact on social outcomes. However, the conditioning of helplessness appears to be incomplete, since residents of long-corridor housing compete and avoid interaction whenever it is likely. Withdrawal responding, associated with learned helplessness, was frequent only when interaction was not likely.

Wortman and Brehm (1975) have presented a model in which loss of control is initially associated with reactance and attempts to reestablish control. As expectations for control diminish with repeated exposure, reactance fades, and behavior becomes more characteristic of helplessness. Given that students arrived on campus expecting to be able to control their social experience, it can be argued that the violation of these expectations aroused reactance among long-corridor residents, and excessive competitiveness may have reflected attempts to reestablish con-

[1]Kurlander *et al.* (cited in Seligman, 1975) found that subjects exposed to solvable problems before playing this kind of game competed frequently and withdrew infrequently. Prior exposure to unsolvable problems, however, led to increased withdrawal and decreased competitiveness, suggesting that the withdrawal response reflects some form of motivational deficit.

trol over social interaction. As these residents became aware of the relatively uncontrollable nature of contact in the dormitories, however, purposive control-oriented behavior seemed to decrease.

Hypotheses

The present research was concerned with the applicability of Wortman and Brehm's (1975) analysis of helplessness to the social dynamics of these dormitory settings. It was hypothesized that long-corridor residents would respond to disconfirmation of expectations regarding regulation of social contact by initially attempting to recover lost control. This reactance was expected to fade as length of exposure (residence) increased, yielding to a more generalized learned helplessness. It was predicted that long-corridor residents would respond to the modified Prisoner's Dilemma (PD) game more competitively after 1 and 3 weeks of residence but that by the end of the 7th week of residence, this competitiveness would decline and withdrawal responding would increase. Short-corridor residents, who presumably do not experience disconfirmation of control expectations, should respond more cooperatively and should not alter their response strategies over time.

It was also predicted that subjects would associate the competitive response option with attempts to discourage interaction and with negative interpersonal goals and that the withdrawal option would be seen as appropriate when desired involvement in the situation was low. Long-corridor residents were expected to identify negative interpersonal goals as indicative of their own through the end of the 3rd week of residence, but by the end of 7 weeks of residence, their goal choice would reflect reduced involvement and motivation. Short-corridor residents were expected to identify social and game-related goals as indicative of their own.

Finally, it was predicted that response to a detailed questionnaire administered to an independently drawn sample of long- and short-corridor residents would provide data paralleling performance in the experimental situation. By asking questions about residents' feelings toward their neighbors, roommates, living conditions, and general college experience, we expected to find long-corridor residents expressing negative feelings toward college life and beginning to complain about difficulty in regulating social contact following 1 and 3 weeks of residence. By the end of the 7th week, it was predicted that interpersonal affect would become more neutral, as affect more characteristic of helplessness emerged, and that problems related to social control would intensify. Again, short-corridor residents' responses were not expected to change over time and were expected to be more positive than those of long-corridor residents. Verification of these predictions would provide support for the model of helplessness proposed by Wortman and Brehm (1975), as well as provide further evidence of a link between crowding and learned helplessness.

The Study

Subjects

A total of 120 freshmen residents of long- and short-corridor dormitories, randomly assigned to place of residence, participated in this study. Sixty subjects from each dormitory design were selected from housing lists and this sampling was random, with the exception that equal numbers of men and women were required. Of these, 15 men and 15 women from each dormitory were selected for participation in the experimental phase of the study. The remaining 60 subjects were asked to complete questionnaires as participants in a survey of campus life.

The experimental sessions were conducted by a college-age male or female experimenter, each conducting half of the sessions in each condition. When female subjects participated, a college-age woman served as a confederate posing as a subject, and when male subjects were considered, a college-age male assumed this role. Neither experimenters nor confederates were aware of experimental predictions or subjects' residence. The questionnaire was administered by a college-age woman.

Design

Residence (long- and short-corridor dormitory) and length of residence (1, 3, or 7 weeks) were crossed with sex of subject in a 2 × 2 × 3 factorial design. Groups of 10 experimental and 10 survey subjects from each dormitory type were tested at each interval. Samples were drawn so that each subject participated in only one mode and at only one testing time.

Three major dependent variables were assessed during the experimental session: frequency of competitive, cooperative, and withdrawal responding in a modified PD game situation. Additional information about goals during the game was obtained by having subjects complete a brief questionnaire at the end of the experimental game.

Survey subjects answered 42 questions about their satisfaction with campus life and dormitory life, the ways in which they spent their time, problems associated with dormitory living, their perceptions of crowding, feelings about roommates, and their motivation to assume control over certain situations.

Experimental Procedures

At the end of the 1st, 3rd, and 7th weeks of dormitory residence, those freshmen residents selected for the first, second, or third testing period were called and asked to participate in a two-person bargaining game. Subjects arrived alone and encountered the experimenter seated at a large table in the experimental room. The room was unfurnished except for the table, three chairs, and two large wooden enclosures housing the game materials. The enclosures consisted of three plywood boards (1 × .5 m) attached by hinges so that they could "wrap around" subjects seated inside them. A game matrix was attached to each center panel, and envelopes containing response cards were attached below the matrix. An envelope in which subjects placed their response on each trial was placed on the table within the enclosures.

After greeting the subject, the experimenter reminded him/her that the session would involve playing a bargaining game with another student who had not yet arrived. Subjects were then seated inside the enclosure on the far side of the room. Once seated inside the enclosure, subjects could not see the experimenter or the confederate, who arrived about 1 minute later.

After the confederate had been greeted and seated inside the second enclosure, the experimenter explained that they would be playing a 20-trial game and asked participants not to talk during the game. Participants were told that the object of the game was simply to score points; those scoring in the highest 10% of all subjects would share a bonus. Subjects were not told to maximize the differences between their score and their partner's, nor were they asked to cooperate. Rather, the possibility of both players in each session winning a postexperiment prize was made clear without providing a strategy for playing the game.

Participants were then told that on each trial they were to select a response from those available (triangle, circle, or square), place it in their game envelope, and pass it on to the experimenter. The experimenter would then read the two choices aloud and determine the outcomes, as participants followed along on the matrices provided. All outcomes depicted in these matrices were explained, and instruction continued until subjects understood how the game was played. Once ready, participants played three practice rounds and 20 game trials, each lasting about 1 minute. Following completion of the 20 game trials and a brief questionnaire, subjects were debriefed, paid, and thanked.

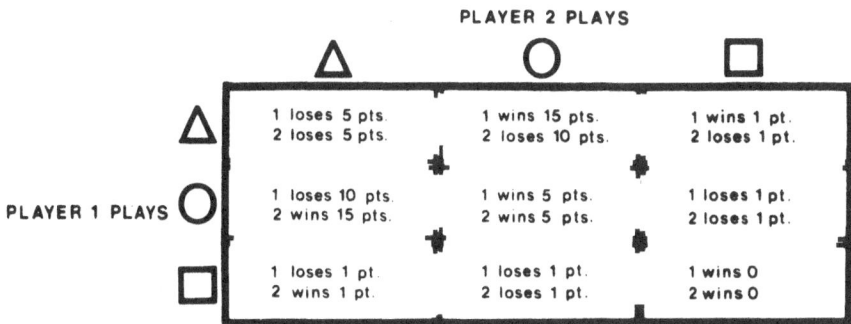

Figure 1. Prisoner's Dilemma game matrix (from Baum & Valins, 1977).

The experimental game was a modified, three-choice Prisoner's Dilemma game similar to that used by Kurlander *et al.* (cited in Seligman, 1975). On any trial, a player could make any of three responses reflecting competitive (triangle), cooperative (circle), and withdrawal (square) strategies (see Figure 1). If both players chose to compete, each lost 5 points, and if both chose to cooperate, each won 5 points. When one chose to compete while the other cooperated, the competer won heavily (15 points) and the cooperative player lost heavily (10 points). If either player chose to withdraw, outcomes were reduced, with players either winning or losing 1 point or breaking even. Withdrawal was viewed as most reflective of helplessness, since it may have represented an inability to arrive at decisions or an unwillingness to participate in a purposeful way.

During the game, confederates played a tit-for-tat strategy (e.g., Sell, 1976); on a given trial, the confederate's response choice was always the same as the response selected by the subject on the preceding trial. The confederate's first play during the practice segment was a triangle, and all subsequent choices reflected the subject's strategy with a one-trial delay.

The primary behavioral indexes were subjects' response choices during the 20-trial game. Following completion of the game, subjects were asked to identify a strategy (playing mostly triangles, circles, or squares) that would be best and most suited for a series of goals, including desires to communicate negative affect to the other player, to win as many points as possible, to discourage the other player from talking to them, not caring what happened in the game, wanting to get to know the other player, and wanting both players to win as many points as possible. Finally, these subjects were asked to indicate which, if any, of these goals reflected the goals they had played for during the game.

Survey Procedures

After the 1st, 3rd, and 7th week of dormitory residence, those freshmen selected for the first, second, or third survey phase were contacted and asked to complete a questionnaire dealing with their college experiences. The survey administration periods corresponded to the three experimental testing periods. Questionnaires were administered in the subjects' bedrooms, and those participating did not receive compensation for their time.

The questionnaire consisted of 41 7-point scales assessing students' feelings about the college, faculty, other students, their roommates, their dormitories, and how they spent their time. Another item presented subjects with a list of 27 general problems (e.g., noise, people in hallway, poor lighting, lack of privacy) and asked them to indicate which, if any, bothered them during the preceding week.

Findings

A 2 (residence) × 3 (length of residence) × 3 (game choices) analysis of variance was performed for the experimental game data, treating game choices as repeated measures to account for their interrelatedness. Two × three (Residence × Length of Residence) analyses of variance were performed on most of the survey data. In both instances, sex-of-subject variance was minimal. Data assessed on 7-point (1–7) reference scales are reported as means with higher values, indicating more of the dimension being measured. Subsequent comparison of interaction means was based on the procedures suggested by Tukey (Myers, 1966). Data reflecting subjects' choice of strategies and identification of goals and dormitory problems were treated with a chi-square analysis.

Experimental Game

Predicted interactions between residence and length of residence for game choices were obtained (see Table 1). Frequencies of competitive, cooperative, and withdrawal choices were affected by residence, $F(2,108) = 39.152$, $p < .001$; by length of residence, $F(4,108) = 3.529$, $p < .01$; and by an interaction of the two, $F(4,108) = 2.926$, $p < .05$. Long-corridor residents were generally more competitive and less cooperative than short-corridor residents, and both groups gradually decreased competitiveness. Long-corridor residents maintained low rates of cooperative responding across test times, whereas short-corridor residents increased cooperative responding during the final test period.

Table 1. Mean Percentage of Competitive, Cooperative, and Withdrawal
Responses ($N = 20$)

	Response								
	Competitive			Cooperative			Withdrawal		
				Testing time					
Dormitory design	1	2	3	1	2	3	1	2	3
Long-corridor	68	66	54	18	17	10	14	17	36
Short-corridor	50	48	37	39	39	52	11	13	11

Rates of withdrawal responding were comparable for long- and
short-corridor residents during the first two testings. Short-corridor residents maintained this low response rate, but long-corridor residents
showed a significant increase in withdrawal responding during the final
testing period ($p < .05$).

Game Strategies and Goals. Postexperimental questionnaires
produced data that suggest that competitive responding in the PD game
reflects a desire to communicate negative affect and to avoid social contact and that the withdrawal response, as suggested by Kurlander *et al.*
(cited in Seligman, 1975), is associated with a lack of motivation to participate in the game. When asked to indicate which of the three response
strategies was most appropriate when one's goal in the game was to
show dislike for the other player, 80% of long-corridor and 73% of
short-corridor residents identified the competitive strategy (playing
mostly triangles). Fifty-eight percent identified the competitive response
strategy as appropriate when one wanted to discourage interaction in
the setting, and only 13% indicated that the withdrawal response was
well suited to this goal. When one did not care what happened in the
game, the withdrawal response was deemed appropriate by half the
residents of each dormitory.[2] The competitive response strategy was

[2]Since the meanings of responses in PD games are difficult to determine, these data were
supplemented by contrasts between goals identified by subjects as reflective of their own
motives and their responses during the game. Of most immediate importance is subjects'
interpretation of the withdrawal response; in order to determine whether this choice
reflected reduced motivation in the game or a more purposeful avoidance/escape strategy,
the frequency of long-corridor residents' withdrawal responding during the third testing
was compared with choice of goals. Of those subjects playing above the median rate of
withdrawal responding (7.5 choices), *all* indicated that they did not care what happened
in the game, and only one said that he was trying to avoid interaction as well. Of those
subjects playing the fewest number of withdrawal choices, only two said that they did not
care about the game, and four indicated a desire to avoid interaction.

Table 2. Number of Subjects Identifying Goals as Reflective of Their Own
($N = 10$)

	Goal								
	Show dislike			Discourage interaction			Do not care		
	Testing time								
Dormitory design	1	2	3	1	2	3	1	2	3
Long-corridor	5	6	1	2	0	6	0	5	7
Short-corridor	0	0	0	0	0	0	1	0	0

also associated with winning as many points for oneself as was possible.
Cooperative strategies were associated with wanting to meet the other
player and wanting to win as many points for both players as was
possible.

After considering the suitability of response strategies for these
game goals, subjects indicated which ones reflected their own goals in
the game (see Table 2). Wishing to express dislike for the other player
was more frequently selected by long-corridor residents, and the identification of this goal dropped sharply after 7 weeks of residence (Fisher
exact probability $< .001$).[3] This decrease in identification of desires to
communicate negative affect to the other player paralleled an increase
over time in the identification of players' goals, such as, "Don't really
care what happens in the game" (Fisher $p < .001$). Selection of
avoidance goals (discouraging interaction with the other player) was
made only by long-corridor residents, and this was marked by a sharp
increase during the final test period ($p < .001$). Short-corridor residents'
goals did not change over time, 70% choosing goals reflecting winning
individual points, 40% choosing goals reflecting desire to win points for
the group, and 33% choosing goals reflecting a desire to get to know
their partner.

In summary, long-corridor residents played a noncooperative
strategy, while short-corridor residents approached balance by playing
cooperative and competitive choices nearly equally. As can be seen in
Figure 2, long-corridor residents' strategies varied with length of expo-

---

[3]During postexperimental interviews, most of these subjects qualified their selection of
this goal. Generally, these qualifications were directed toward letting us know that their
dislike for the other player was "nothing personal." Most indicated that they felt negatively about the other player because, as one subject said, "He wouldn't let me play the
game the way I wanted to."

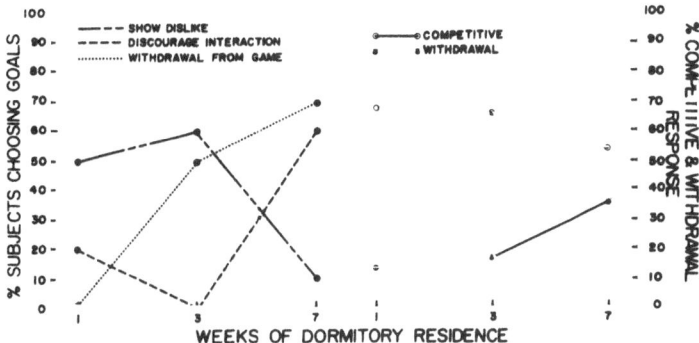

Figure 2. Goal selection and game responding by long-corridor residents.

sure; during the first two test periods, they responded competitively, but by the third, this responding had fallen off in favor of sharply increased withdrawal. Similarly, those goals selected by long-corridor residents reflect their response choices; during the first and second testings they were likely to identify negative interpersonal goals as reflective of their own, but by the final testing, identification of this goal had yielded to reduced motivation.

Survey Findings

Satisfaction. Long-corridor residents were generally less satisfied with the college ($M = 3.9$) than were short-corridor residents ($M = 4.9$), $F(1,54) = 5.291$, $p < .05$. Similarly, long-corridor residents expressed more negative feelings toward other students on campus, $F(1,54) = 4.112$, $p < .05$. These differences were stronger when liking for the freshman class was considered, $F(1,54 = 10.417$, $p < .01$, and were strongest when dormitory and floor neighbors were rated. Paralleling their rate of competitive responding in the experimental game, long-corridor residents reported more negative feelings toward dormitory neighbors, $F(1,54) = 41.670$, $p < .001$, but did not differ from short-corridor residents after 7 weeks of residence, $F(2,54) = 3.407$, $p < .05$ (see Table 3). Further, long-corridor residents reported less attraction for floor neighbors than did short-corridor residents, $F(1,54) = 30.455$, $p < .001$. Although the interaction term for this analysis did not reach significance ($p < .15$), differences between long- and short-corridor residents appear strongest after the 1st week of residence.

Short-corridor residents expressed slightly more negative affect for floor neighbors during the second testing period than in the other two. Mean comparisons, however, indicated that differences among short-corridor residents were not significant.

Table 3. Reported Dislike for Dormitory and Floor Neighbors[a]

	Dislike					
	For dormitory neighbors			For floor neighbors		
	Testing time					
Dormitory design	1	2	3	1	2	3
Long-corridor	3.80	3.70	2.00	4.20	3.00	3.40
Short-corridor	1.40	1.60	1.50	1.80	2.50	1.90

[a]7-point scale: 1 = like very much; 7 = like not at all.

Dormitory Life. While there were no differences along any dimension when considering time spent sleeping, watching television, studying in the library, or socializing outside the dormitory, interesting differences between long- and short-corridor residents were found for some behavior *in* the dormitory. Long-corridor residents reported that they spent less time studying ($M = 3.7$ vs. 5.1) or socializing ($M = 3.6$ vs. 4.7), $F(1,54) = 12.360$, $p < .001$, and $F(1,54) = 10.092$, $p < .01$, respectively. Interestingly, long-corridor residents reported that they spent as much time socializing in their dormitories after the 1st week ($M = 4.6$) as did short-corridor residents ($M = 4.7$), but reported a decrease in this time after the 3rd ($M = 3.6$) and 7th weeks ($M = 2.5$), relative to consistent reports from short-corridor residents ($M = 4.5$ and $M = 5.00$, respectively), $F(2,54) = 3.695$, $p < .05$.

Regulation of Interaction. While both long- and short-corridor residents reported frequent interaction with neighbors in hallway areas of their dormitories, long-corridor residents began to report problems in controlling this interaction as length of residence increased. As can be seen in Table 4, long-corridor residents reported greater difficulty in regulating interaction in the dormitory than did short-corridor residents, $F(1,54) = 11.386$, $p < .01$. However, long-corridor residents did not differ from short-corridor residents along this dimension after the 1st week of residence; significant differences did not emerge until the 3rd week of residence ($p < .05$). Differences in expectations for control were indicated by greater long-corridor students' disagreement with the statement, "It is worthwhile to try to structure your interaction with others (in your dormitory)," $F(1,54) = 10.31$, $p < .01$ (see Table 4). Again, long- and short-corridor residents responded comparably after the 1st week of residence but diverged over time. Although the interaction term for this analysis did not reach significance ($p < .15$), mean comparisons indicated that ratings after 1 week were comparable, ratings after 3 weeks

Table 4. Mean Ratings of Disagreement with Control-Related Statements[a]

Dormitory design	Not worthwhile trying to change things			Easy to control interaction in dormitory			Worthwhile to try to structure interaction in dormitory		
					Testing time				
	1	2	3	1	2	3	1	2	3
Long-corridor	5.40	4.90	3.60	3.10	5.00	4.90	3.80	5.10	4.90
Short-corridor	5.10	5.20	5.50	3.00	3.80	2.50	3.40	3.30	2.60

[a]7-point scale: 1 = agree; 7 = disagree.

were marginally divergent ($p < .06$), and ratings after 7 weeks were significantly different ($p < .01$).

This pattern of results, suggesting that difficulties concerning regulation of interaction increase over time for long-corridor residents, is also evident when considering reports about the frequency of interaction and desires to avoid and withdraw. As can be seen in Table 5, long-corridor residents reported more frequent unwanted interaction than did short-corridor residents, $F(1,54) = 38.616$, and reported more frequent unwanted contact after the 7th week of residence than after the 1st or 3rd weeks, $F(2,54) = 3.316$, $p < .05$. This pattern also emerged when subjects indicated how frequently they ignored people in their dormitories; although long-corridor residents ignored others more often than did short-corridor residents, $F(1,54) = 7.865$, $p < .01$, this difference was generated primarily by divergence after 7 weeks of residence, $F(2,54) = 3.880$, $p < .05$. Further, long-corridor residents expressed a greater desire to avoid others in their dormitory, $F(1,54) = 26.644$, $p < .001$ (see

Table 5. Mean Responses to Statements Concerning How Often Dormitory Residents Avoid Others or Are Forced to Interact When Interaction Is Not Wanted[a]

Dormitory design	Wish to avoid others			Encounter unwanted interaction		
			Testing time			
	1	2	3	1	2	3
Long-corridor	1.50	2.20	4.30	3.00	3.00	4.00
Short-corridor	.80	1.10	1.00	1.20	1.20	1.00

[a]7-point scale: 1 = never; 7 = often.

Table 5), but did so primarily after 7 weeks of residence, $F(2,54) = 5.870$, $p < .01$. Again, responses did not differ after the 1st week of residence, but were quite different after 7 weeks of dormitory living.

Crowding and Helplessness. Long-corridor residents reported that they felt that their dormitory was more crowded ($M = 4.4$) than did short-corridor residents ($M = 1.6$), $F(1,54) = 86.122$, $p < .001$. While long-corridor residents showed a slight tendency to feel more crowded over time, and short-corridor residents felt less crowded over time, this interaction effect did not reach significance ($p < .20$). Even though problems that have been associated with residential crowding (i.e., unwanted interaction, difficulty regulating social contact) were not salient for these students until the 3rd or 7th week of residence, long-corridor residents felt crowded in their dormitories almost immediately.

Findings concerning residents' feelings of control outside the dormitory provided equivocal support for the notion that helplessness begins to emerge with increasing length of residence. Although the results did not reach significance, long-corridor residents were more likely to agree with the statement, "It is often not worth the effort to try to change the way things are," than were short-corridor residents, especially after the 7th week of residence, $F(2,54) = 3.054$, $p < .075$. Long-corridor residents also showed a tendency to disagree with the statement, "In the library it is worthwhile to try to stake out a space and defend it," more strongly than did short-corridor residents, but were more likely to do so after 7 weeks of residence, $F(2,54) = 2.75$, $p < .08$.

Identification of Problems. Finally, subjects were presented with a list of 27 potential dormitory problems and were asked to indicate which had bothered them during the preceding week. Overall, long-corridor residents reported more problems ($M = 6.0$) than did short-corridor residents ($M = 3.0$), $F(1,54) = 25.298$, $p < .001$. An interaction between residence and length of exposure to dormitory conditions indicated that short-corridor residents reported more problems after 3 and 7 weeks of residence than after the 1st week, $F(2,54) = 4.083$, $p < .05$. Considering types of problems reported, short-corridor residents were most likely to complain about the physical atmosphere of the dormitory (e.g., noise, condition of bedroom, lighting), whereas long-corridor residents were more apt to identify problems relating to privacy or social dynamics (e.g., people in the hallway, too many conversations to remember, too many people you don't really know). Long-corridor residents had more difficulty maintaining adequate privacy than did short-corridor residents, $F(1,54) = 5.15$, $p < .05$. Similarly, long-corridor residents reported more problems with social contact in the hallways than did short-corridor residents, $F(1,54) = 63.24$, $p < .001$ (see Table 6). Mean comparisons indicated that long-corridor residents reported sig-

Table 6.   Number of Problems with Social
Interaction and Sharing Common
Facilities Identified by Subjects

	Social interaction[a]			Common facilities[b]		
	Testing time					
Dormitory design	1	2	3	1	2	3
Long-corridor	17	26	26	4	3	3
Short-corridor	4	8	5	1	6	3

[a] People in hallway, too many conversations to remember; possible 50/cell.
[b] Availability of washing facilities, crowded bathrooms and lounges; possible 40/cell.

nificantly more social problems after 3 and 7 weeks than after 1 week of residence ($p < .05$). Unexpectedly, long-corridor residents did not report many problems with crowded bathrooms and other shared areas, whereas short-corridor residents did (see Table 6). While these items do not reflect overload or reduction of residents' ability to regulate social contact, they do suggest that short-corridor residents also experienced some difficulty adjusting to dormitory life.

Conclusions

This research was concerned with assessing the utility of control-based analysis of crowding by examining the relationship between prolonged exposure to high social density and motivational deficits characteristic of learned helplessness. Previous research has suggested that crowding and helplessness are linked (e.g., Baum & Valins, 1977; Rodin, 1976), and it was predicted that loss of regulatory control over social interaction due to large residential group size would condition a variant of helplessness. This process was seen as being mediated by expectations for control. Consistent with Wortman and Brehm's (1975) description of helplessness conditioning, it was predicted that initial recognition of uncontrollable social outcomes would arouse negative interpersonal affect and generate attempts to restore control. As expectations for control diminished as a function of increased exposure to uncontrolled residential conditions, helpless responding was expected to increase.

The data generally confirm these expectations. High social density and large residential group size were once again associated with crowding; long-corridor residents reported more crowding, frequent unwanted interaction, less satisfaction, and more problems than did short-corridor residents. Further, these complaints were associated with reductions in perceived control, as long-corridor residents also reported greater difficulty regulating dormitory contacts, lower expectations for such control, and desires to avoid and ignore neighbors. These control-relevant problems became more salient as length of residence increased, and long- and short-corridor resident behavior diverged as control became a problem in the long-corridor housing. Crowding experienced as a function of sharing space with large numbers of others appears to be associated with loss of the ability to regulate interpersonal contact and with threats to personal control.

Cognitive Mediation of Helplessness

The data also suggest that, while withdrawal and rather purposive avoidance behavior are modal responses to prolonged exposure to crowding stress, helplessness is also conditioned. Further, they indicate that helplessness is mediated by expectation for control, emerging only after repeated attempts to reestablish control have proven unsuccessful. Although short-corridor residents showed few effects of length of exposure to residential conditions, long-corridor residents reported an intensification of control-relevant problems in their dormitory as length of residence increased. Their responses to the experimental game and survey items show trends that suggest dramatic shifts in mood and behavior as these problems become more salient.

During the first 3 weeks of residence, long-corridor residents responded to the game competitively, selected negative interpersonal goals, and reported negative feelings abut neighbors, other students, and college life in general. By the end of 7 weeks of residence, both competitive responding and the expression of negative interpersonal affect had decreased. Paralleling the intensification of control-related problems, expectations of control dropped after 3 weeks of residence, and long-corridor residents began to show reduced involvement in the experimental game. Four weeks later, competitive game choices had decreased as withdrawal choices increased; negative interpersonal game goals had been replaced by reduced motivation and desires to discourage interaction with the other player; and reported helplessness increased. These changes were also accompanied by increased motivation to avoid other people and by a decrease in the amount of time spent in the dormitory setting.

This pattern of response closely approximates the sequential process of helplessness conditioning described by Wortman and Brehm (1975) and further suggests that control is a relevant dynamic of the experience of social density. Responses through the 3rd week of residence are suggestive of reactance; long-corridor residents expressed negative affect during this period, and if competitive responding is considered as an attempt to control the game, this behavior may be reflective of generalized motivation to maintain control in interpersonal situations. By the end of the 3rd week of residence, however, long-corridor residents appeared to recognize that they could not control their social experience in their dormitories, and they withdrew from the dormitory. By the end of 7 weeks of residence, expectations for control were low, and behavior was more characteristic of withdrawal and helplessness.

On the surface, long-corridor residents' continued competitiveness and increasing desire to avoid interaction seem inconsistent with a helplessness interpretation. Avoidance is clearly purposeful, and coping strategies directed toward discouraging interaction are not indicative of helplessness. One could argue for the primacy of well-learned avoidance responding over helplessness, in which helpless responding is manifest only when social interaction is unlikely, but a closer inspection of the data suggest an alternative explanation. By comparing subjects' goals during the game with their responses during the game, it becomes apparent that some of these subjects were trying to avoid interaction, while others were responding helplessly. Of the 10 long-corridor residents considered after 7 weeks of residence, only 3 expressed desires to avoid contact and reduced motivation in the game, and these subjects averaged 6.7 withdrawal responses and 10.3 competitive responses during the game. Subjects who expressed reduced motivation in the game but did not express desires to avoid contact averaged 9 withdrawal responses and 9.5 competitive responses; those wishing to avoid contact but not indicating reduced motivation averaged 5 withdrawal and 13.7 competitive responses during the game. The most competitive subjects were those interested in avoiding interaction, and their low rate of withdrawal responding is consistent with their failure to report reduced motivation in the game. Those subjects who did report reduced motivation but were not interested in avoiding contact were less competitive and played more withdrawal choices. Apparently, some subjects exhibited symptoms of helplessness, whereas others were more purposeful in game response. The identification of factors that differentiate among these response styles is an interesting topic for future research.

Development of Helplessness in Dense Settings

These findings provide support for Wortman and Brehm's (1975) model of helplessness suggesting that naturalistic threats to personal control, when experienced repeatedly and when apparently unresolvable, are associated with sequential phases of reactance and helplessness. They also suggest that crowding reflects control-threatening aspects of high density and lend support to control-based models of crowding (e.g., Baron & Rodin, 1978). Dormitory experience described as "crowded" was also characterized by density-mediated threats to social regulation, intensified by architectural arrangement of interior space. Changes in environmental potential (from what was expected) led to relatively stable but ineffective coping responses, and appraisal of the responses' success seemed to result in resignation to loss of personal control. Both the conditioning of helplessness and the persistence of potentially maladaptive avoidance strategies may be seen as long-term costs of crowding stress and underscore the significance of control-based intervention strategies when dealing with high density settings (e.g., Langer & Saegert, 1977; Sherrod & Cohen, Chapter 13, this volume). By viewing crowding as the stressful loss of personal control due to high density conditions, the consequences of high density may be better predicted and their impact may be reduced.

Acknowledgments

The authors would like to thank Kathryn Maye, Lisa Passalacqua, Winthrop Piper, and William Shoff for their help in conducting this research. We would also like to thank Carlene S. Baum, Glenn E. Davis, Ilene Gochman, and Camille B. Wortman for their helpful readings of this article.

References

Aiello, J., Epstein, Y., & Karlin, R. The effects of crowding on electrodermal activity. *Sociological Symposium*, 1975, 14, 43–57.

Altman, I. *The environment and social behavior*. Monterey, California: Brooks/Cole, 1975.

Baron, R., & Rodin, J. Personal control and crowding stress: Processes mediating the impact of spatial and social density. In A. Baum, J. Singer, & S. Valins (Eds.), *Advances in environmental psychology* (Vol. 1). Hillsdale, New Jersey: Erlbaum, 1978.

Baum, A., & Koman, S. Differential response to anticipated crowding: Psychological effects of social and spatial density. *Journal of Personality and Social Psychology*, 1976, 34, 526–536.

Baum, A., & Valins, S. *Architecture and social behavior: Psychological studies of social density*. Hillsdale, New Jersey: Erlbaum, 1977.

Desor, J. Toward a psychological theory of crowding. *Journal of Personality and Social Psychology*, 1972, 21, 79–83.

Freedman, J. *Crowding and behavior*. San Francisco: Freeman, 1975.

Glass, D., & Singer, J. *Urban stress: Experiments on noise and social stressors*. New York: Academic Press, 1972.

Langer, E., & Saegert, S. Crowding and cognitive control. *Journal of Personality and Social Psychology*, 1977, 35, 175–182.

Myers, J. *Fundamentals of experimental design*. Boston: Allyn & Bacon, 1966.

Rodin, J. Crowding, perceived choice, and response to controllable and uncontrollable outcomes. *Journal of Experimental Social Psychology*, 1976, *12*, 564–578.

Saegert, S. High density environments: Their personal and social consequences. In A. Baum & Y. Epstein (Eds.), *Human response to crowding*. Hillsdale, New Jersey: Erlbaum, 1978.

Schopler, J., & Stockdale, J. An interference analysis of crowding. *Environmental Psychology and Nonverbal Behavior*, 1977, *1*, 81–88.

Seligman, M. *Helplessness*. San Francisco: Freeman, 1975.

Sell, R. *Cooperation and competition as a function of residential environment, consequences of game strategy, and perceived control*. Unpublished doctoral dissertation, State University of New York at Stony Brook, 1976.

Sherrod, D., & Cohen, S. Density, personal control, and design. In J. Aiello & A. Baum (Eds.), *Residential crowding and design*. New York: Plenum Press, 1979.

Stokols, D. A social-psychological model of human crowding phenomena. *Journal of the American Institute of Planners*, 1972, *38*, 72–84.

Stokols, D. The experience of crowding in primary and secondary environments. *Environment and Behavior*, 1976, *8*, 49–86.

Sundstrom, E. Toward an interpersonal model of crowding. *Sociological Symposium*, 1975, *14*.

Worchel, S., & Teddlie, C. The experience of crowding: A two-factor theory. *Journal of Personality and Social Psychology*, 1976, *34*, 30–40.

Wortman, C., & Brehm, J. Responses to uncontrollable outcomes: An integration of reactance theory and the learned helplessness model. In L. Berkowitz (Ed.), *Advances in experimental social psychology* (Vol. 8). New York: Academic Press, 1975.

Crowding and Residential Design

John R. Aiello and Andrew Baum

Introduction

The past decade has witnessed the development of a small but rapidly growing body of literature that has a high degree of relevance and potential applicability to residential environments. It should be possible to use much of this information for the purposes of (1) reducing or eliminating the consequences of life in high density settings and (2) altering or designing environments that facilitate positive experiences of their residents rather than causing these residents difficulties that they have to overcome.

The research reported in Part I of this book indicates that one of the distinguishing characteristics of high-density residential environments is the set of potentially negative consequences associated with feelings of loss of control over unwanted social contacts in these environments. As Freedman (Chapter 10) suggests, high density may actually serve to intensify the negative effects of social isolation and anonymity often found in urban environments. Newman's (1972) research further illustrates how design problems in secondary environments (e.g., hallways) can create unsafe residential conditions and greater fearfulness among residents. It should be possible to enrich residential experiences so that social stressors are brought under control by applying what is known about the spatial needs and requirements of individuals and by employing architectural designs more congruent with these needs.

Considerable evidence indicates that people actively use space and manipulate the physical environment to regulate social interaction (see Altman, 1975). We suggest that intervention strategies involving a combination of design modifications and personal, social, and situational processes be employed to provide residents with environments that are conducive to greater opportunities to exert more effective control in these primary settings. It should also be possible to include ongoing training and information regarding the more effective use of existing residential spaces. Useful sources of information are available from accounts of successful coping strategies employed by people of other cultures living under high-density conditions (see Aiello & Thompson, in press). For example, Michelson (1970) has reported:

> The Japanese exemplify successful adjustment to very high densities. Faced with huge urban masses in a country with no room in which to expand, and without the precedents for high rise construction, the Japanese have made their dwellings small, and private open space is minimal. The Japanese have reacted to this pressure by "turning inward." They strongly distinguish between what is private and what is public in physical as well as social terms. Interiors of homes are personal, and their lack of size is compensated for by an intensity of detail. Every inch is open space for utilization through physically undifferentiated use of interior space. Every room may potentially be used like any other, with only the movement of portable partitions as prerequisite. (p. 155)

Given the complexity of environment–behavior relationships, we believe that it is both necessary and desirable to have a good deal of collaboration among professionals in diverse fields. In his analysis of public housing constructed in the United States for low-income families, Hall (1966) has noted:

> To solve formidable urban problems, there is the need not only for the usual coterie of experts—city planners, architects, engineers of all types, economists, law enforcement specialists, traffic and transportation experts, educators, lawyers, social workers, and political scientists—but for a number of new experts. Psychologists, anthropologists, and ethologists are seldom, if ever, prominently featured as permanent members of city planning departments but they should be. (p. 169)

While we agree with Hall's design team approach, potential problems need to be acknowledged. Because practitioners of environmental design (e.g., architects, planners, urban designers) and behavioral and social researchers have very different orientations to the resolution of environmental issues, interdisciplinary communication can often be quite difficult. Only through these cooperative efforts, however, will we be able to solve problems concerning residential crowding and design. It is hoped that the issues presented in this book will contribute to the momentum of these collabora‥ve problem-solving efforts.

Jonathan L. Freedman, who pioneered some of the early experimental work on human crowding, provides in Chapter 10 a series of implications for housing design based on his examination of the current status of crowding research. After a brief reflection upon confusion between crowding as a psychological experience and as a physical variable, he reports that he still finds questions related to the latter more basic and more interesting. Unlike Galle and Gove, McCarthy and Saegert, Holahan and Wilcox, and others in this volume, Freedman feels that when social and economic factors are controlled, high residential density is *not* usually harmful. Rather, high-density levels are viewed to be simply more suitable for some people and for some situations than for others. Similarly, the effects of a particular architectural design are most dependent on situational and populational characteristics. He suggests that the focus of crowding research should be on (a) the delineation of the "complex and subtle effects" that high-density housing exerts on the way people feel and behave, and (b) the examination of when housing works and when it does not. While acknowledging the failures of high-rise housing design for low-income families, he cautions that the lack of residential choice may be more important than architectural factors. He proposes that (a) everyone, including low-income families, should be given some choice of where to live, and (b) low-income projects should allot more money for amenities, careful planning, and maintenance. In concluding, Freedman notes that architectural design can sometimes play an important role in residential satisfaction and suggests that long corridors be eliminated in high-density housing by breaking them into smaller units.

A data-based analysis of the architectural impact of the long-corridor design conducted by the second editor along with Glenn Davis and Stuart Valins is presented in Chapter 11. In a long-term research program conducted on two college campuses, homogeneous student populations residing in traditional long-corridor college dormitories were compared with those who lived in four- to six-person suites or short-corridor dormitories. Survey and interview procedures were used to formulate hypotheses, which were subsequently tested using a mixed strategy of field, laboratory/experimental, and observational methodologies. Survey data indicated that corridor residents felt more crowded and overwhelmed by unwanted and uncontrollable interaction than did suite residents, and observation of behavior in the dormitory environments suggested that corridor residents were more likely to avoid social contact with their neighbors. Unobtrusive observations in a waiting room also indicated that long-corridor residents looked less, interacted less frequently, and chose seats more distant from other people. Taken together, the findings of this program of research provide evidence that design factors not only mediate the quality of residential life but are associated with persistent social stress and social withdrawal in other settings as well. This chapter concludes with a discussion of a site-intervention that has been implemented in an effort to improve the quality of life in the long-corridor environment. Consistent with Freedman's analysis, the floor of each long-corridor has been subdivided and lounge areas provided to each cluster of residents, thereby reducing group size and providing semiprivate interaction zones. It is believed that this intervention will enhance the individual resident's ability to regulate social contact and provide an opportunity to further test the conceptual linkages between architecture and behavior.

In Chapter 12, a somewhat different approach for the application of human spatial research to the design and planning process is taken by Gary W. Evans. Unlike many other researchers who have focused on the microenvironment, Evans considers the implications of spatial research for the design of the microenvironment (immediate, interior spaces). In his review of the psychological and physical variables that impact on spatial behavior he discusses two general areas of personal space and crowding research that should be helpful to the designer: (1) investigations that have specified "comfortable" spatial requirements as a function of individual characteristics, interpersonal situations, and social settings; and (2) the small number of empirical studies that have manipulated design parameters and measured the impact on perceived crowding. A brief synopsis of the dominant theoretical perspectives in human spatial behavior is presented, wherein the construct of stress is posited to be the critical mediating link. Although Evans discusses the limitations of the existing body of proxemic research, some suggestions for future work are also advanced. He proposes that an alternative research process in which researchers and designers collaborate during the design process should be adopted in order to address the problems of setting specificity and limited temporal sampling in human spatial research. More specifically, hypotheses involving design and program variables should be tested in existing as well as proposed real-world settings, providing the opportunity both for prospective investigations and for postconstructive user evaluations.

Sherrod and Cohen (Chapter 13) propose that controllability may be the single most important factor that mediates the effects of crowding. They argue that distinguishing between controllable and uncontrollable high-density environments may, in fact, be the key to understanding the circumstances under which density adversely affects behavior. According to Sherrod and Cohen density has one of two possible effects: (1) increasing unpredictability and therefore the uncontrollability of a given environment, or (2) restricting freedom and constraining behavior, producing a sense of helplessness. Throughout their chapter, they suggest that the effects of crowding depend upon the behaviors being performed in an environment as well as the individual's perception of the environment as controllable or uncontrollable. The belief in one's ability to control the environment is interpreted as not necessarily implying the ability to actually implement control. Consistent with this view, they postulate that the environments cannot only be physically designed to be more controllable but modified as well, through social and cognitive interventions, so as to appear more controllable. Among the ways suggested to increase the predictability and controllability of environments are: (1) creating spaces within apartments that allow a retreat from social interaction; (2) building smaller apartment houses, breaking up long corridors, and decreasing the number of people served by a given recreational environment to increase predictability, control, and friendship formation; and (3) defining neighborhood landmarks, points of interests, and other salient features in urban environments to enhance perceptions of control. Issues encountered in designing more controlled environments are also discussed, along with the idea that the cognitive environment may be more important in the determination of human behavior than the physical environment.

In the final chapter, Schiffenbauer discusses some of the problems confronted in attempting to identify housing designs that will ameliorate negative effects resulting from high-density living. He suggests that the most critical research question that needs to be addressed by psychologists and architects alike is: What conditions make high density living tolerable? As Schiffenbauer acknowledges, however, successful collaboration between the two disciplines is often difficult. He notes that the differences in orientation held by psychologists and architects (e.g., applied vs. basic) often do not mesh. Moreover, he indicates that those architects who have suggested that high-density designs can be quite pleasant have often failed to consider the needs of individuals who will be inhabiting buildings or to conduct systematic evaluations concerning the effects of buildings on the behavior of residents. The results of two investigations are reported that show some of the negative effects of high-density living may be controlled through the manipulation of architectural factors. More specifically, satisfactory high-density living conditions may be created either by physically designing spaces so that they are perceived as more spacious and uncrowded or by providing residents with the skills necessary to regulate their social interactions. The importance of the issue of control is also discussed.

References

Aiello, J. R., & Thompson, D. E. Personal space, crowding, and spatial behavior in a cultural context. In I. Altman, J. F. Wohlwill, & A. Rapoport (Eds.), *Human behavior and environment* (Vol. 4), *Culture and environment*. New York: Plenum, in press.

Altman, I. *The environment and social behavior*. Monterey, California: Brooks/Cole, 1975.

Hall, E. T. *The hidden dimension*. New York: Doubleday, 1966.

Michelson, W. *Man and his urban environment: A sociological approach*. Reading, Massachusetts: Addison-Wesley, 1970.

Newman, O. *Defensible space: Crime prevention through urban design*. New York: Macmillan, 1972.

10

Current Status of Work on Crowding and Suggestions for Housing Design

Jonathan L. Freedman

The population explosion has given birth to an explosion of a different sort—an enormous outpouring of research on the effects of crowding. Ten years ago there were virtually no papers by psychologists on this subject. Five or six years ago the first ones started appearing and since then there has been an exponential increase in their number. Surely the last few years have seen more papers on crowding than appeared in all the time before that. Now that this area has become a recognized topic for research and has attracted a great many psychologists, perhaps it is time to stop for a moment and see what we know now. This is particularly appropriate since it seems to me that we have passed the initial stage of research—laying out some of the problems and discovering the most obvious facts about crowding—and it is now time to enter the second phase, which should consist of studying the more complex issues and explaining some of the early results.

First, I think by now we have overcome an initial confusion caused by various meanings of the term *crowding*. When I first started my research, I assumed that psychologists and especially environmental psychologists would be concerned primarily with crowding defined entirely as a physical variable. Accordingly, I devoted all my time to investigating the effects of density—the amount of space available per person. Although this seemed quite clear to me, it turned out that some people in the field were more interested in crowding as a psychological feeling

Jonathan L. Freedman • Department of Psychology, Columbia University, New York, New York 10027.

or state, and, unfortunately, the two meanings of the term became confused. Stokols (1972) and others tried to make clear the distinction between crowding as a physical variable and crowding as an emotional response, but for some years people still talked about one when it seemed as if they meant the other, and the terms were being used interchangeably.

I hope that by now all this has been clarified. The terms refer to two different issues: one is how people respond to the amount of space that is available to them and the other is how people respond when they feel cramped or when they feel they do not have enough space. Obviously, the two issues are related but are quite distinct. My own preference, and I think that of most people in the field, is to study the former question because it is more basic and more interesting. Once people feel crowded, virtually by definition this is a negative state, and, presumably, they will respond in ways people always respond to negative states. As far as I can tell, there is nothing that distinguishes this negative emotional state from others, and, therefore, it is not of special interest to the environmental psychologist. In any case, people are obviously free to choose which question they would like to study, and the only important point is to keep the questions separated. In this chapter, I shall be talking about crowding as a physical variable.

Second, the initial finding from research on crowding is that it is not clearly harmful to human beings. Research in the laboratory, in field settings, and in the real world of cities and housing indicates that by and large people are not negatively affected by living or playing under conditions of high density. The laboratory research occasionally produces a negative effect of high density, but of the dozens of studies that I know of on this topic, very few have found main effects of density. This is even more true of real-world research in which, with one or two exceptions, no studies have found any appreciable relationships between density and pathology once the obvious economic and social factors are controlled. The work by Mitchell (1971), and Galle, Gove, and McPherson (1972), as well as ours (Freedman, Heshka, & Levy, 1975), make this fairly clear.

Studies of high-density housing by Michelson (1973) and Wellman and Whitaker (1974) have found exactly the same thing. Indeed, by now the fact that high density is not generally negative is so taken for granted by sociologists that they seem to find the question either trivial or naive (e.g., Fischer, 1976). All the summaries of research in this area have come to this same conclusion: Whatever else it may do, high density is not generally harmful to people.

Now let us be clear that this does not by any means imply that high density is *never* harmful. The point is that the effects of density depend entirely on the situation and on the people involved. It is not a stressor in the usual sense. Pain, exceedingly loud noise, and severe hunger are always stressful and always have harmful effect on the individuals involved. Crowding is not always stressful and does not always have harmful effects. Under most circumstances, high density does not produce arousal and does not affect performance (Freedman, Klevansky, & Ehrlich, 1971, and many other studies). But, under appropriate circumstances, it does produce physiological arousal and does at least occasionally interfere with performance on complex tasks (Aiello, Epstein, & Karlin, 1975). This study, which shows physiological arousal, was conducted in an extremely small room (4 feet by 4 feet) with six subjects who were told that the purpose of the study was to assess their response to the room and who engaged in no social interaction. Under these conditions, high density is arousing and does produce the expected physiological responses. That is perfectly fine and certainly consistent with all of the other research in the field. It demonstrates that it is possible to construct a situation in which high density is arousing. Exactly what conditions are necessary to produce such arousal is not entirely clear. We do know that a wide range of conditions with quite high levels of density do not generally produce arousal. Perhaps very intense crowding is necessary; perhaps the lack of social interaction is necessary; and perhaps the subjects' attention must be drawn to the size of the room. All these are guesses—we have no idea what the crucial factors are. All we do know is that crowding does sometimes produce arousal and sometimes does not. Indeed, that is where we stand at the moment—crowding is not usually arousing or harmful to people, but the response depends on the situation.

Third, the main point of all the research is that the effects depend on the situation. The sex of the subjects, whether or not the groups are mixed, why the people think they are there, whether or not there is social interaction, and the quality of that interaction all seem to affect people's response to high density. Males generally seem to respond more negatively than females, but that is not entirely consistent. Such nonsocial behaviors as task performance seem to be relatively unaffected by high density, but that too is not entirely consistent. Situations that are inherently pleasant or positive in any way seem to be made more pleasant by high density, whereas negative situations seem to be made more negative. Yet the evidence for this is not overly convincing, and I imagine that it too is not entirely consistent. Finally, it seems likely that the arbitrariness of the situation and the reason people are given for being in a crowded condition play a role.

These diverse findings leave us in need of some explanations of the effects of crowding on people. A number of so-called theories have been offered; in fact, almost every week we seem to see another theory of crowding being suggested even though almost none of the ones that I have seen tried very hard to demonstrate that they can explain all the data. It seems clear that none of the existing theories handles all the data and that calling them theories is dignifying them more than they deserve. Basically, what we have is a number of hunches or small-scale hypotheses that attempt to bring some order out of the seeming chaos of findings on crowding. Milgram's notion (1970) of sensory overload explains some of the results, especially the existence of inconsistency. Although he does not carry the idea this far, one could use his notion to explain why some people do well in cities and others do not, why males and females have different responses to high density, and even, if you wanted to stretch the notion, why the same person will respond positively under one condition and negatively under another. Unfortunately, there is virtually no independent support for the phenomenon of sensory overload, and at the moment this framework does not make any specific predictions about the effect of crowding.

The theory is much more useful if it is combined with the idea that every individual has an optimal level of stimulation, and that some like very high levels of stimulation whereas others like low levels. Once you assume this, many inconsistencies make sense, and it also explains why some people like cities and others dislike them. In other words, not only would it explain the lack of overall harmful effects of crowding but it could also account for some of the positive effects and some of the negative effects. Without this, sensory overload should never lead to a positive effect of high density, and since we get as many positive effects as negative ones, the theory clearly would be wanting.

My own notion (Freedman, 1975) that high density intensifies typical reactions by making other people in a room a more important stimulus is also a minitheory at best. It was devised to account for the inconsistent effects and especially for the sex difference that people generally find. It is very difficult for me to accept the fact that males and females would differ in their response to a variable as basic as the amount of space they have, and, therefore, I looked around for some other explanation. There is some research that specifically supports the idea that, at least under some circumstances, high density will magnify or intensify people's reaction to a situation. When conditions were explicitly made positive, both males and females responded more positively under high than under low density; and the opposite was true when the situation was explicitly made unpleasant. However, the evidence is still quite scanty, and, in any case, I am sure that this is only part of the explanation.

Another fact that almost certainly plays a role is how we interpret the situation and what level of density we consider appropriate. A number of authors have suggested that we should explain crowding effects in terms of attribution, because how we respond depends on how we interpret the situation. Personally, although attribution theory has been applied to many psychological phenomena, I think it has only limited reference to the effects of crowding. Appropriateness is certainly important; attribution processes may enter in occasionally, but not as one of the basic factors. However, let us regard that as one possible explanatory principle.

I have not tried to list all the explanations of crowding that have been proposed. There are a number of others, some of which will probably turn out to be useful and some not so useful. Personally, I think it is premature to try to devise a complete theory of crowding, or even to hope to explain all the data we have available at the moment. There are still too many inconsistencies and too many basic facts we do not know. I think the focus from now on should be on specifying what conditions produce what effects. Let us accept the fact that crowding is not always harmful or always beneficial to people, that we do not have an instinctive territorial response or anything like that, and that sometimes we like to be under conditions of high density and sometimes we dislike it. Having accepted that, and understanding that our response depends on the particular situation, let us now try to describe those situations in detail. Once we have a fairly large body of data showing which condition produces which effects, we can try to come up with a theory that explains all our diverse findings.

Although we are still at an early stage of research on crowding, I think we can say something about the effects of high-density housing. We have good reason to believe that the density of housing will not be an important factor in rates of pathology, social or otherwise. We already know that high-density cities and neighborhoods do not produce any more crime or mental disturbance than low-density areas as long as economic factors are equated. Moreover, even the amount of space available in the home is unrelated to pathology. Thus, the question should not be whether high-density housing is positive or negative but rather how its effects on people differ from those of low-density housing. In other words, we should not state the question in terms of good or bad. Instead, we can start looking for complex and probably subtle effects on the way people feel and behave.

The same is true of high-rise housing. Although there are those who condemn such housing as being unfit for human life, obviously a great many people live in high-rise apartments and seem to function just fine. Indeed, many people who can afford to live anywhere in any kind of housing choose to live in high-rise apartments and even pay a premium to live on the higher floors rather than live close to the ground. There is a considerable amount of research showing few if any differences between residents of high- and low-rise housing (e.g., Michelson, 1973; Wellman and Whitaker 1974).

All this is consistent with what we know from studies of crowding. High density has not been shown to have generally harmful effects, and, therefore, we should expect no overall negative effects of either high-density or high-rise living. This is an important point, because there sometimes seems to be the feeling that high density and high rise are inherently bad and that any lack of negative effects are due to people's ability to adapt to this basically bad situation. If you start with this view, of course, you will view high-density living as a problem that must be overcome. In contrast, if you accept my view (and that of most of the research) that high density is not inherently harmful, you can then look at housing and try to decide what kind would be best for the particular circumstances under consideration. Neither high- nor low-density housing is better than the other, but each may be more suitable for some people and some situations.

The difficulty is that at the moment we do not know the specific effects of the various kinds of housing (if there are any), and, therefore, we are not in a good position to make suggestions. We do know that some high-density, high-rise housing is eminently successful in the sense that people choose to live in it, continue to live in it, and seem to be quite happy with it. But we also know that some housing of this sort is a disaster, with people avoiding it, moving out if they get a chance, destroying their own building, and so on. Some high-rise housing has deteriorated to the extent that it has become uninhabitable and is eventually deliberately destroyed—rather than being left as a monument to poor planning or whatever the reason is for failure.

Our job is to discover when housing works and when it does not. One obvious starting point is that the failures of high-rise housing are, as far as I know, always with buildings designed for low-income families. Typically, these people have little or no choice about where to live—they are assigned to the housing and cannot afford to say no. This lack of choice may be more important than any architectural feature—if you must live somewhere, you may start off with a bad attitude and not take pride in your home. Clearly, the ideal would be to offer everyone, including low-income families, some choice of where to live rather than forcing them to live in high-rise buildings (or low-rise for that matter, although there seems to be less resistance to that).

A second point about these low-income projects is that they usually allocate less money for amenities, for careful planning, and perhaps most important, for maintenance. The basic design of the building may matter much less than factors such as how often the halls are cleaned, the efficiency and reliability of the elevators, hot water and heating systems, and probably most crucially, the security system. We should not make the mistake of blaming the design or the density level until we make certain the fault does not lie in these other areas.

Thus, I do think that design can sometimes play an important role. Although I have nothing in principle against either high-density or high-rise buildings, I think the combination of the two generally puts some strain on the residents in certain specific ways. The social milieu of a large building is quite different from that of a smaller building. In particular, someone who shares a corridor with 29 other families and rides in an elevator that services 30 such floors comes into contact with many people. This makes it difficult to get to know all the faces and especially difficult to become friendly with anyone. Thus, at least for a while, a resident of such a building feels surrounded by strangers rather than neighbors. Some of us may prefer that—it preserves our privacy through anonymity—but it prevents a sense of community and certainly reduces the security of the building because strangers are less likely to be recognized as such.

As I have suggested elsewhere, one way of overcoming these effects might be to eliminate the long corridors that are so often a feature of high-rise buildings. You can keep the same number of floors and apartments per floor but break up the corridors into smaller units. This may require more elevators, but if it works, it will probably be worth the cost. In any case, the idea is that each family would then share a corridor with only a few other families (perhaps five) and would be more likely to get to know their neighbors. This would tend to produce a sense of community on the floor. If each corridor unit had its own elevator, this would also greatly reduce the number of different people using the elevators and might facilitate getting to know people on other floors. In other words, instead of the 900 families in the building (30 per floor, 30 floors) all sharing the same facilities, they would be broken up into groups of 6 per floor and 180 per elevator. Although 180 families is still a large number of people, it might be a manageable number at least for recognizing faces. All this is highly speculative—we have no data to show that it would work or indeed have any effect at all. But the idea is consistent with previous research in the field and seems plausible.

In general, our approach to housing should start with the assumption that the effect of a particular design or type of building will depend on the situation, just as the effect of density does. And our focus should be on trying to pick the design that will be most appropriate for a given situation and population. We do not yet know exactly how to do this, but I think we have some plausible ideas, and they should be tried. In the future, perhaps we will have done enough research so that we have solid answers. For now, we should base our suggestions on the evidence we do have and maybe they will be helpful.

References

Aiello, J. R., Epstein, Y. M., & Karlin, R. A. Effects of crowding on electrodermal activity. *Sociological Symposium*, 1975, *14*, 43–57.

Fischer, C. S. *The urban experience*. New York: Harcourt Brace, 1976.

Freedman, J. L. *Crowding and behavior*. New York: Viking, 1975.

Freedman, J. L., Heshka, S., & Levy, A. Population density and pathology: Is there a relationship? *Journal of Experimental Social Psychology*, 1975, *11*, 539–552.

Freedman, J., Klevansky, S., & Ehrlich, P. The effect of crowding on human task performance. *Journal of Applied Social Psychology*, 1971, *1*, 7–25.

Galle, O. R., Gove, W. R., & McPherson, J. M. Population density and pathology: What are the relations for man. *Science*, 1972, *176*, 23–30.

Michelson, W. *Environmental change*. Research Paper No. 60, Center for Urban and Community studies, University of Toronto, 1973.

Milgram, S. The experience of living in cities. *Science*, 1970, *167*, 1461–1468.

Mitchell, R. E. Some social implications of high density housing. *American Sociological Review*, 1971, *36*.

Stokols, D. On the distinction between density and crowding: Some implications for future research. *Psychological Review*, 1972, *79*, 275–277.

Wellman, B., & Whitaker, M. High-rise, low-rise: The effects of high density living. Ministry of State, Urban Affairs Canada, 1974, B.74.29.

11

Generating Behavioral Data for the Design Process

Andrew Baum, Glenn E. Davis, and Stuart Valins

When behavioral scientists seek to study architectural impact or become involved in the design process, they are often unfamiliar with architectural methods and with many of the problems faced by designers. Architects rely upon scientific theories, methods, and data to construct stable physical structures and upon past experience when considering aesthetic components of design, but they typically use less systematic assessments of user perceptions and needs. User populations may not be specified in great detail when designs are conceived, and survey-based assessment of those populations that are known may not allow identification of design variables that cannot be easily articulated. The designer is required to approximate the needs and expectations of people with whom contact is unlikely and must translate clients' estimates of needs and spatial requirements into minimally expensive, maximally effective environments. The number of decisions that must be made, combined with the pressures of limited resources, nonspecific client needs, and estimates of space use, and a lack of data about the effects of specific design variables or configurations of variables have led to continued reliance on intuitive judgments of behavioral impact. The results of these decisions may be positive, but the legacy of Pruitt-Igoe (e.g., Yancey, 1972) serves to remind us of the unexpected results of

Andrew Baum • Department of Medical Psychology, Uniformed Services University, School of Medicine, Bethesda, Maryland 20014. Glenn E. Davis • Department of Psychology, Washington College, Chestertown, Maryland 21620. Stuart Valins • Department of Psychology, State University of New York, Stony Brook, New York 11790.

such decisions. Clearly, a thorough and systematic consideration of the social and psychological effects of various design features must precede the development of future environments.

The collection of design-relevant behavioral data has been approached in different ways, most notably through direct questioning or observation (e.g., McCarthy & Saegert, Chapter 4, this volume; Newman, 1972; Wolfe, 1975). By studying relatively homogeneous populations in different housing through wide-based surveys, McCarthy and Saegert revealed powerful correlates of high density and housing type. Despite strong and consistent relationships among these variables and indices of social overload and pathology, the nonexperimental nature of these studies has caused some to approach their findings with caution. Similarly, observational studies have provided substantive data but have been criticized for the necessary reduction of experimental control. Without debating the relative merits of different research strategies (see Winkel, 1976, for a detailed discussion of their issues), we wish to present an alternative approach to the assessment of architectural impact that combines the strengths of survey, observational, and quasi-experimental research strategies.

Basic to our perspective is the comparison of relatively homogeneous and comparable user populations of different environments that can be contrasted along specific design variables. In an experimental sense, treatments are the direct result of architectural variation and the assessment of these treatments is conducted in much the same way as in the laboratory. For example, if residents of two different college dormitories have been randomly or arbitrarily assigned to residences and are demonstrated to be comparable before living in the dormitories, differences that emerge in behaviors in the dormitory and in other settings can be interpreted in terms of the differential impact of their residential environments. Because the architectural treatment (e.g., high-rise vs. low-rise; corridor vs. suite) is independent of the study of it in the fullest sense of the word, attempts to understand the treatment effects should yield reliable data regarding responses to specified environmental conditions.

By studying homogeneous student populations residing in architecturally different dormitories on two college campuses, we have been able to observe the effects of design variables that moderate perceptions of others and produce varying levels of social density (e.g., Baum & Koman, 1976). By using survey and interview procedures to formulate hypotheses and by testing these speculations using field and laboratory experimentation and observation, we were able to isolate, with a respectable degree of specificity, those design variables associated with persistent social stress and social withdrawal. Although we acknowledge the uniqueness of both our settings and our user populations and are aware of subsequent limits on generalization to other situations, we have pursued this research for three reasons. First, we have revealed relationships that can be used in the design of future dormitorylike environments. Second, we have followed, in a somewhat logical fashion, a multilevel analysis of existing environments and have demonstrated an interesting way of obtaining information about response to design features. Finally, corroborative or supportive findings in other settings (with different populations) have suggested that, although limited in immediate use, our work in the dormitories has identified conceptual linkages between architecture and experience which can be used to explain behavior in other settings. In this spirit, then, we shall summarize our program of research in college dormitories.

The Stony Brook Research Program

Our study of the effects of built environments on user populations has been a long-term endeavor. Like other areas of applied social research, the research problem was identified before the research team arrived on the scene, and the statement of this problem (i.e., how to design residential environments in the face of population growth and inevitable high-density conditions) did not suggest immediate, empirically testable hypotheses. A lack of behavioral data-based analyses of architectural impact made it apparent that it would be necessary to consider information from many sources to identify links between built environments and human behavior.

At the outset, two sources of empirical information about population density, social organization, and pathology were readily available: the work of several animal ecologists (e.g., Calhoun, 1962, 1967; Christian, Flyger, & Davis, 1960), and that of urban researchers (e.g., Schmitt, 1963; Galle, Gove, & McPherson, 1972). Feeling that the environment–behavior processes in the dormitories might be conceptualized in ways analogous to those used by animal ecologists, we focused our attention on a detailed description of the ecological features of these residences. Stated design intentions of the architects and our own opinions guided us in the development of naive hypotheses of how the architectural and social systems of dormitories might interact. Surveying residents' perceptions and sytematic observation of user behavior in the dormitories helped us specify residential social patterns associated with different designs. Inputs from these sources facilitated the derivation of hypotheses that could be experimentally tested.

Generally speaking, the development of the research program has proceeded from a descriptive level to a point where social patterns *in situ* and differential sociability in other settings can be explained as consequences of the manner in which different designs engender interpersonal contact and facilitate social expectations. Although it is clear that recommendations for design interventions have and can be made on the basis of intuition and descriptive analysis, the experimental orientation we have used has yielded empirical relationships of a causal nature. The ability to ascertain causal relationships between the built environment and behavior represents an advance beyond descriptive analysis and appears to be the first step necessary for the development of a theory of behavioral design.

When we first considered the problem of high-density residential design, our primary sources of information were our intuitions and the pioneering studies of high-density animal populations conducted by Calhoun (e.g., 1962). Calhoun found that population growth resulting in increasing group size was associated with, among other things, severe social dysfunction and reproductive failure (1962, 1970). Increasing numbers of rodents using the same resources, even when these resources were *not* scarce, resulted in increasing numbers of "collisions" with other animals and made it more difficult to regulate social contact. Calhoun felt that, in order to maintain a satisfying balance between gratifying and frustrating social encounters, these contacts had to be controlled. As population density increased, regulation of social interaction became more difficult, and rather severe pathology resulted.

Of course, Calhoun was working with rats and mice. Even though other investigators had found similar pathologies among highly concentrated populations of other animals (e.g., Christian, 1963), generalizations to humans were considered with a great deal of caution. We believed that Calhoun's analysis was applicable to humans but felt that the mediation of density would be more complex and the consequences less severe. Our only sources of data for high-density human populations were descriptive and rather inconsistent analyses of the relationship between population density and pathology (e.g., Galle *et al.*, 1972; Schmitt, 1966), and a few conflicting experimental treatments of these relationships (e.g., Freedman, Klevansky, & Ehrlich, 1971; Griffitt & Veitch, 1971).

Milgram's (1970) description of social overload in the city and Glass and Singer's (1972) analysis of the effects of noise-induced stress provided direction for our speculations and suggested applications of Calhoun's ideas. Control was clearly an important variable, and loss of control appeared to be stressful. In addition, loss of control over interactions was viewed as being associated with social withdrawal and un-

satisfactory interpersonal relationships. Convinced that regulation of social interaction was important for people as well as for animals, we
began to consider design alternatives that would minimize or maximize
the control-debilitating effects of high density. Based on Stokols's (1972)
description of mediation of density, we felt that some architectural designs would result in loss of social control under high-density conditions, whereas others would not. In effect, then, we were searching for
analogues of Calhoun's crowded rat universe.

Remembering that Calhoun was able to distribute crowding by the
arrangement of spaces and connecting access ramps, we began to consider the dormitories housing undergraduates on campus. For our purposes, the homogeneous subject population and rather arbitrary housing assignment process were ideal. Because we knew that we would be
investigating the effects of phenomena over which we could exert no
direct experimental control, it was important that subject variance be
kept minimal. After considering floor plans and observing dormitories
during the summer (while empty) and during the fall (while full), we
settled on two designs characterizing 10 buildings. One design structured each floor into a single large group and was therefore seen as
maximizing the stressful consequences of high density. The other, housing students in smaller 4- to 6-person groups, was not expected to cause
crowding and stress.

The Dormitory Designs

Of the two designs studied, the corridor design was more common.
Dominated by a long, undifferentiated central hallway, these dormitories housed 34 residents on a floor. However, there were no groupings or subdivisions of this large group beyond that afforded by the
two-person bedroom unit. Instead, 17 double-occupancy bedrooms, a
central bath area, and an end-hall lounge were arranged along a
double-loaded corridor. Residents shared these living areas, and the
hallway provided social space as well as access to facilities and to the rest
of the building (see Figure 1).

Housed in comparable physical densities, the 34 or 36 residents of a
floor in the suite-design dormitories were not required to share many
resources with all 33 others. Instead, the design of these residential
environments dispersed people in small, 4- to 6-person suites, each
containing its own bath and lounge. Thus, the only area shared by the
larger group in these buildings was the hallway space, and, because
shared facilities were *inside* the suites, the hallway space did not provide
access to living spaces shared by all. Suite units arranged along this

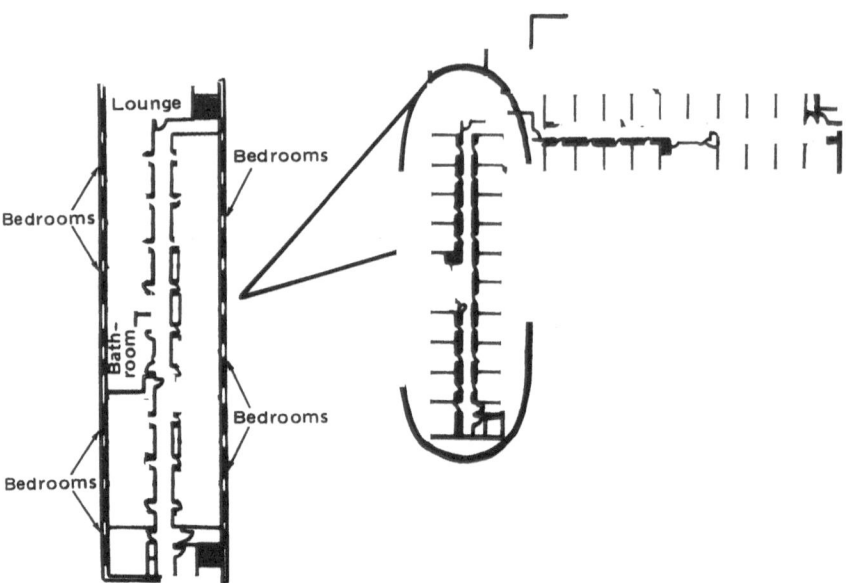

Figure 1. Floorplan of corridor-design dormitory housing. From Baum & Valins (1977). Reprinted with the permission of Lawrence Erlbaum Associates.

hallway were relatively self-sufficient, and use of the hallway seemed to be restricted to access to other suites and to the rest of the building (see Figure 2).

Considering Calhoun's work and Milgram's description of urban overload, we felt that the large number of residents sharing living space in the corridor dormitories would result in their "colliding" with each other more frequently. This, we felt, would lead to conditions characterized by (1) interaction in the hallway when interaction was inappropriate or unwanted, (2) interaction with others not known or with whom comfortable modes of interaction had not yet evolved, and (3) a general loss of control over interaction in this hallway space. If, for example, a student wished to use the bathroom, he would have to leave his room and walk through the hallway to get there. Because 33 other residents also used this space, the likelihood that others would be in the hallway was relatively high. Thus, regardless of their desire for social contact, residents were likely to encounter others as soon as they left their bedrooms.

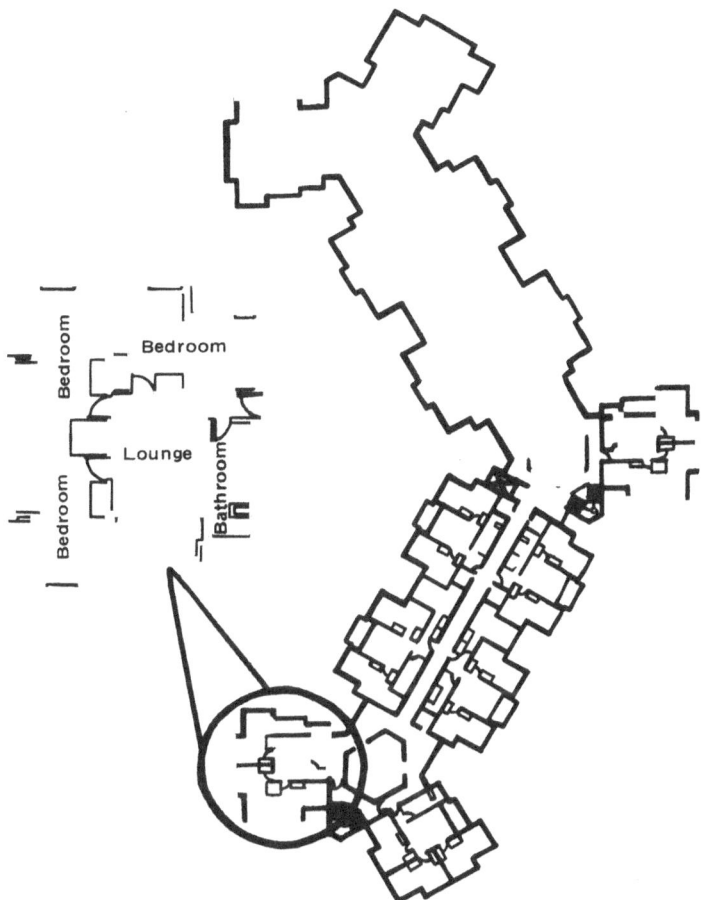

Bedroom

Bedroom

Bedroom

Bedroom

Lounge

Bathroom

Figure 2. Floorplan of suite-design dormitory housing. From Baum & Valins (1977). Reprinted with the permission of Lawrence Erlbaum Associates.

Although our initial analyses were somewhat more specific, their general implications regarding the nature, frequency, and control of social interaction are sufficient to illustrate our first attempt at applying these theoretical constructs to real environments. We knew that the physical densities in the two dormitory complexes were comparable and obtained extensive background data which suggested that residents of corridor and suite housing were comparable prior to entering college residence halls. We were also satisfied that the dormitory designs were generally comparable; furnishings were similar, surroundings were not very different, and the ratio of upper-class to freshman residents was nearly identical on the floors studied. The primary difference between the designs was the way in which they arranged interior space and distributed social resources on the floor. The next step was to assess resident experience in these dormitories and determine the accuracy of our speculations.

Social Experience: "I Feel Crowded"

This initial research phase relied heavily on self-report by residents of the dormitories. In these and most other studies, freshmen residents were considered because the relative impact of dormitory living was presumed to be greater on newcomers to these settings. Students were surveyed and observed in the dormitory settings, and findings were used to generate additional assessments and experimental hypotheses. Generally, our initial predictions were verified. Of greater importance, however, these findings amplified our understanding of the social dynamics of these environments, suggesting complexities overlooked during our initial theorizing.

Of initial interest, these investigations revealed that, as expected, corridor residents were likely to feel crowded in their dormitory while suite residents were not. Even though these students were exposed to comparable physical densities, their ratings of crowding were consistently divergent. Furthermore, these feelings seemed to be related to the frequency of unwanted interaction reported. Ratings of crowding in corridor dormitories varied with the location of the bedroom on the floor; residents living closest to the areas where most people congregated or that were used more frequently (e.g., bathrooms, lounge) felt *more* crowded than did residents living at the far end of the hall. Corridor residents were also more likely to indicate that there were too many people living on their floor, that there were others they would like to avoid, and that the rate of interaction in their dormitories was too high. Corridor residents were not satisfied with the amount of privacy provided, and reported higher frequencies of interactions that were unwanted or inconvenient. While suite residents were vaguely positive about their residential experience, corridor residents reported that they felt crowded, complained about the frequency and control of social interaction in the dormitories, and generally indicated negative residential experience.

These findings suggested that Calhoun's ideas about group size and crowding were applicable to people. Furthermore, they indicated that design variables that varied group size, even when physical density remained constant, were influential in determining response to design and density. However, we had only survey data upon which to base our hypotheses; we still had no evidence that this syndrome of crowding was stressful or was associated with pathology. Although we were more confident in our description of the impact of these dormitory designs on mood, we needed evidence of behavioral differences in order to extend our analysis.

During the period of time in which we began considering experimental tests of these findings, we conducted behavioral mapping of the dormitories. A variant of controlled observation techniques, behavioral mapping provided us with a record of behaviors that occurred in these environments and identified their location. By mapping behavior on three corridor and three suite floors, we obtained initial evidence of different behavior in the two settings.

The results of this effort provided evidence of social withdrawal and suggested an unexpected link with Calhoun's (1967) discussions of *social velocity*. Most instances of social behavior occurred in the hallways of the corridor dormitories and in the small lounges in suite-style housing. Interaction was most likely to occur in areas adjacent to bedroom units, but these areas provided different amounts of control and privacy in the different dormitories. When nonsocial behavior was considered, corridor residents were most frequently observed in their bedrooms and suite residents were most often found in the suite lounge. Suite residents continued to use the lounge where interaction was very likely, whereas corridor residents returned to their bedrooms where interaction was considerably less likely. By avoiding those spaces in which interaction was most likely to occur, corridor residents may have been minimizing the frequency of uncontrollable hallway contact. This finding was consistent with corridor residents' reported desire to avoid social contact, and an inverse relationship between actual and preferred location of social encounter among these students further suggested that corridor-design residents were avoiding social interaction.

An alternative explanation of these findings, that corridor residents did not use their hallway space for nonsocial activities because they were less appropriate for things that people do when they are alone, led to further speculation. Yancey (1972) and Newman (1972) had described the impact of unsuitable or nondefensible spaces in residential environments, and it seemed clear that the long, undifferentiated hallways in the corridor-design dormitories were too public to be converted to semiprivate space in which local social control could be exerted and recognized. The suite lounges, on the other hand, were physically removed from the public hallways and were interposed between these spaces and private bedroom areas. Access to the lounge was limited, and the small group clustered around it seemed to be able to use the lounge and to exert control over who used it and when different uses were appropriate. Furthermore, control over interpersonal contact seemed greater in these lounges than in the corridor hallways. Thus, the amount of social control "available" to residents in primary areas for social encounter and neighboring varied by design.

These observations led us to consider expansion of our initial interpretation and suggested our first experimental study. Frequent and unwanted social interaction in the corridor-design dormitories was viewed as being exacerbated by the lack of control available to the group in the hallway spaces where interaction was most likely. Corridor residents were not choosing to interact in these areas but seemed forced to do so by the arrangement of living space. Suite residents, on the other hand, were interacting with others in areas of their choice, and the features of the suite lounge provided the opportunity for group control over these shared spaces. An experimental study was designed to test some of these hypotheses, assessing the relationship between sensitivity to numbers of people and control provided by the situation.

In this initial experiment, we used the model-room technique discussed by Desor (1972). Although these procedures were largely exploratory and not intended to be definitive, we felt that we could obtain valid projective indices of crowding and sensitivity to others. The settings that we created lacked most nonvisual cues that would be present, thereby exaggerating the visual aspects of the situations described. However, for our purposes, these additional cues were not crucial.

One hundred freshman residents of the corridor and suite-style dormitories were visited in their bedrooms and presented with three scale-model rooms (1 inch = 4.3 feet) representing a bedroom, a lounge, and a library reference room. Subjects were asked to place as many miniature people into each room as possible, "before you would feel crowded." The assumption underlying this index was that subjects would arrive at a "threshold of crowding" in which the addition of another figure would cause the room to be perceived as crowded. The activities in the rooms were described so that the bedroom represented a space in which social interaction was very likely but proprietorship provided some control, the lounge represented a room where interaction was again likely but control was not, and the library served as a room where interaction was unlikely and external control high. It was expected that corridor and suite responses would diverge most when control was lacking and interaction was probable (lounge) and would diverge least when interaction was unlikely and control externally supplied by implicit behavioral norms.

The results of this and of subsequent model-room experiments provided general confirmation of these predictions. Unexpectedly, corridor residents did not respond to the control provided by the bedroom space;

they placed fewer figures in the bedroom ($\bar{X}$ = 6.2) and lounge ($\bar{X}$ = 20.2) than did suite residents ($\bar{X}_B$ = 9.4; $\bar{X}_L$ = 26.9). The expected reversal in the library room was confirmed; corridor residents placed a few more figures ($\bar{X}$ = 12.1) in this room than did suite residents ($\bar{X}$ = 11.2).

The Probability of Interaction

While these studies revealed differences in sensitivity to numbers of people as a function of dormitory residence, the degree of control afforded by the situation seemed to influence sensitivity only when extreme comparisons were made. When control was not provided, corridor residents placed fewer figures than did suite residents. When control was externally supplied, corridor residents placed the same number of figures as did suite residents. However, when self-generated control was implied by proprietorship of one's own bedroom, corridor residents again placed fewer figures. These findings may have been caused by a number of factors. General feelings of helplessness involving resignation to the fact that one can rarely control interaction may have made corridor residents less sensitive to control cues (e.g., Rodin & Baum, 1978), causing them to respond only to rather strong descriptions of control. Alternatively, the rather limited nature of model-room procedures may have restricted realism to the point that control in the bedroom was not considered by subjects.

A third explanation is that the probability of interaction in the settings overwhelmed the effects of control provided by the nature of the space. Control provided by these rooms is largely normative and often subtle. The effects of excessive interaction, on the other hand, appear to be more pervasive. Both the bedroom and lounge models were presented in ways that made social interaction in them very likely, while the library model description implied relatively low levels of social interaction. It seemed possible that the effects of frequent and uncontrolled social contact, or the consequences of prolonged exposure to these conditions, were sufficiently strong to render interaction regulation through territorial control less effective. In this case, corridor residents should respond negatively to increasingly likely social contact in settings other than those in which they experience unwanted social interaction, and externally supplied control or social structure should ameliorate this negativity.

In order to assess the relative strength and persistence of the impact of high social density and associated phenomena such as frequent and unregulated interaction, a series of laboratory studies were conducted. The scenarios in these studies were all similar. Subjects were asked to wait, either alone or with a confederate posing as a subject, while an experiment was being readied. By placing subjects in a waiting room and observing their responses to the setting that we had created, we were able to assess some of the persistent effects of prolonged exposure to high social density.

These procedures allowed us to manipulate the probability of interaction during the waiting period. In one study, subjects waited by themselves or with another person; it was predicted that corridor and suite residents would respond comparably when alone, as no interaction was possible. However, when waiting with another person, the increased likelihood of social encounter should cause residents' behaviors to diverge. Unlike suite residents, corridor residents would seek to avoid social involvement.

The results of these experiments provided evidence of this phenomenon. When waiting alone, corridor and suite residents sat in comparable places, looked around the room, and felt relatively comfortable. When asked to wait with another "subject," however, corridor residents sat farther from the other person, looked at him less often, and felt considerably more uncomfortable than did suite residents. This divergence was attributable to increased distancing and discomfort among corridor residents, as suite residents did not respond differently to the two conditions.

A second study manipulated the probability of interaction by asking subjects to wait with a confederate under involvement-enhancing cooperative expectations or involvement-inhibiting competitive expectations. When expectation of involvement with the confederate was low, corridor and suite residents responded in a similar manner. When involvement was more likely, corridor residents again responded to the increased probability of interaction by exhibiting avoidance behavior and reporting unrelieved stress. Suite residents did not respond differently in the two conditions.

All the studies we conducted using these procedures produced the same pattern of findings. Regardless of the nature of the manipulation, corridor residents avoided interaction and reported more discomfort when contact was likely. When magazines were provided and subjects could avoid contact by "burying" themselves in a magazine, corridor residents felt comfortable and did not exhibit avoidance responses. When the magazines were removed, the incidence of avoidance behavior and stress increased significantly among corridor subjects.

These findings argued strongly for the analysis that we had applied to these environments. Architectural differences between the two residential styles mediated constant physical density and resulted in reported crowding and more negative social experience in corridor dormitories. Different *social* densities created by varied grouping of residents around shared resources led to differential sensitivity to the probability of interaction and, apparently, determined the value of social contact with strangers.

From this point, we returned to the survey instrument and began to apply what we had already learned to other concerns. Subsequent efforts to determine whether corridor and suite residents were able to use the control provided by private or semiprivate space suggested that, in a limited sense, they were. Other studies were addressed to the potential consequences of chronic loss of control and indicated that when interaction was unlikely and avoidance was not a competing response, corridor residents exhibited symptoms of learned helplessness (Seligman, 1975).

Some of our most interesting findings grew out of a questionnaire administered to sophomore residents of the corridor- and suite-design dormitories. All these residents had lived in corridor-style housing during their freshman year and had selected their sophomore year dormitory. As a result, we could no longer be certain that subject variance was minimal. However, answers to our questions were provocative. Those living in the corridor dormitories reported that they had tried harder than those who had moved into suites to structure the social climate of their floors by living with a group of friends. However, corridor residents did not report greater cohesiveness on their floors, suggesting that their attempts to structure their residential environment were not as successful as in the suite-design housing. Of greater interest was the finding that corridor residents who reported that they were members of a cohesive residential group did not feel as crowded as corridor residents who were not members of a local group. This effect was observed on floors where some residents felt crowded and others did not.

These findings led to subsequent study that confirmed the relative absence of local friendship groups in the corridor dormitories and suggested that neighbors in these buildings found it difficult to work together. Additional work focusing on social control provided by the groups forming in the dormitories suggested that when these groups did form, they were clearly recognized by others and exerted territorial control over their shared space (Baum, Mapp, & Davis, 1977). Data indicated that when residents were able to get together, the group provided norms and structures that reinforced members' abilities to regulate social contact. The resiliency of the group "membrane" and well-established channels of communication helped group members circumvent unwanted or unpredictable interaction with residents not in the group and to regulate contact within the group.

These findings, considered in light of the underlying architectural manipulation, suggest that design variables facilitating the development of small residential groups will help mitigate the effects of high density.

Throughout an extensive series of studies, analyses, and reanalyses, we eventually arrived back where we had started: *group size*. The crowding and discomfort created by the corridor design seemed due to the large group sharing common living spaces, and the absence of these effects in suite housing was associated with relatively small residential groupings. It is possible, however, for corridor residents to adapt, to substitute a group of their own selection for one created by the physical setting. The development of groups in the corridor dormitories is considerably more difficult than in the suites, but when they do evolve, they appear to serve the same functions as the naturally forming suite groups. This suggests one answer to the question of what to do with high density residential environments. If the interior design of the corridor-style dormitories had been more supportive of local group development, it is possible that the problems we have observed would have been avoided. Thus, the design of high density residential environments should be sensitive to phenomena that influence various forms of group development.

A Conceptual Replication

Despite the apparent robustness of our findings at the Stony Brook research site, we were aware of possible limits to the generalizability of the phenomenon itself and of our analysis. These college residents were primarily from metropolitan New York, and many of them had been able to choose in which dormitory they wished to live. Although self-selection factors did not appear to significantly alter the basic findings, we felt that in order to ascertain their generality, we would need additional data from a different population exposed to a similar architectural manipulation of residential group size. Both the demographic characteristics of students and design of college housing at Trinity College fulfilled our requirements for assessing the generality of our analysis.

Although both Stony Brook and Trinity populations were of comparable age, only about 15% of Trinity residents had lived in metropolitan New York, whereas nearly 80% of the former group had. In addition, reported family income was higher among the Trinity population. Of equal or greater importance, however, was the fact that Trinity students were randomly assigned by the college to their places of residence. Therefore, contamination or confounding of our data because of students' housing preference could be categorically excluded. It seemed clear that if our analysis of Trinity's residential environments yielded data consistent with those generated in New York, the group size in-

terpretation could be used to predict residential behavior in other settings.

The design of Trinity dormitories was less variable than those previously studied, as all residential buildings were of low rise (two–three
floors), double-loaded corridor construction. However, as before, the
dormitories could be differentiated according to the number of residents
clustered around shared facilities and circulation areas. Long-corridor
dormitories housed approximately 36–40 students per floor in double-
and single-occupancy bedrooms along an undifferentiated central hallway. Each floor contained two community bathrooms, but other than
halls and bedrooms, no provisions had been made for areas in which
residents could casually socialize with neighbors. Short-corridor buildings housed about 60 students on a floor in single- and double-
occupancy bedrooms but subdivided residents into three groups of
20–22 students each. Each subsection of a floor was supplied with a
lounge and a bath on the central hallway (see Figures 3 and 4).

Although we were not working with designs that were identical to
those employed by the Stony Brook architects, it was immediately apparent that the consequences of these designs at Trinity were very similar. We had no reason to believe that the suite design was inherently
superior to the short-corridor style, nor did we feel that residential
groups of 4–6 were necessarily more positive than groups of 20 residents
created by short corridors. In fact, there was no way to compare experi-

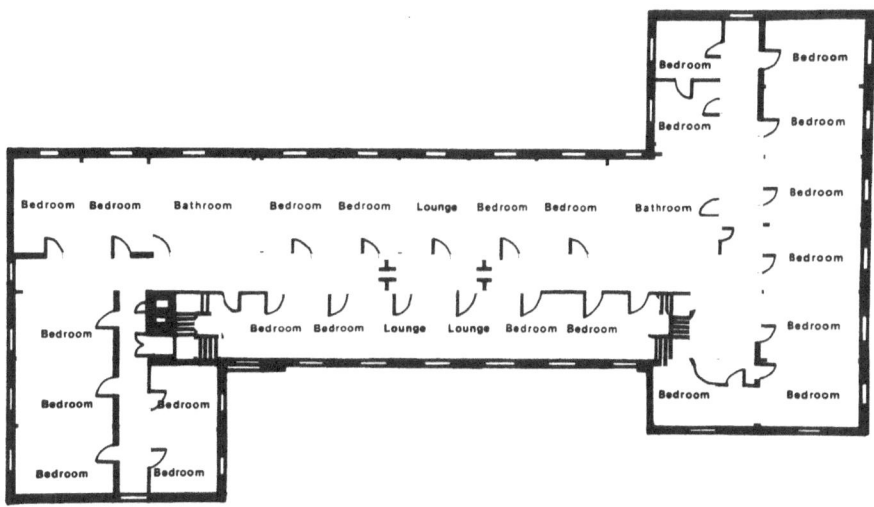

Figure 3. Floorplan of long-corridor dormitory housing. From Baum & Valins
(1977). Reprinted with the permission of Lawrence Erlbaum Associates.

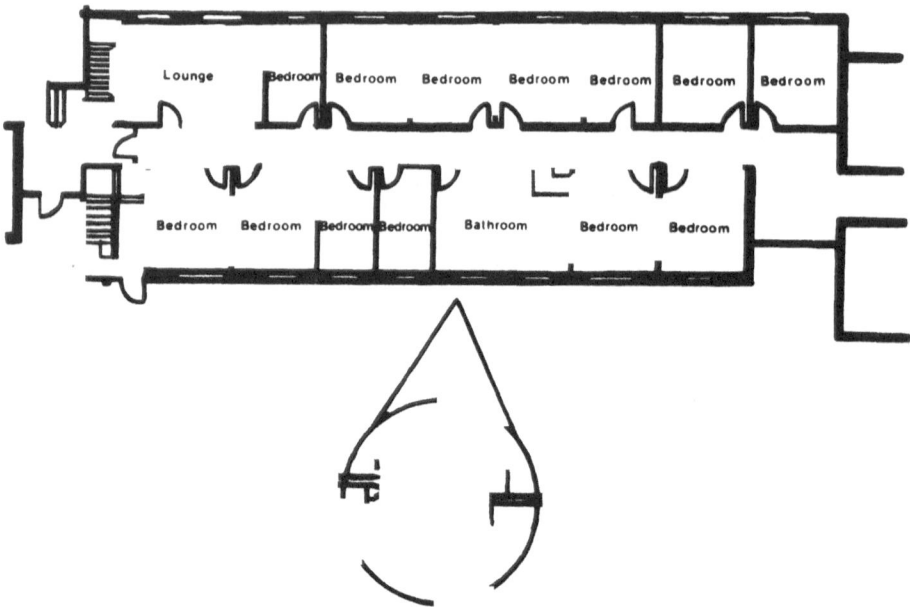

Figure 4. Floorplan of short-corridor dormitory housing. From Baum & Valins (1977). Reprinted with the permission of Lawrence Erlbaum Associates.

mentally the Stony Brook and Trinity designs. However, if we again discovered stable and predictable differences between the residential experiences and social behavior of short- and long-corridor residents at Trinity, our original analysis of the consequences of residential group size would gain considerable credence.

As before, our investigation began by interviewing students about the nature and quality of their local residential experiences. Consistent with our expectations, long-corridor residents reported that their dormitory was more crowded, that they experienced a greater frequency of unwanted social interactions with neighbors, and that they would like to avoid others residing on their floor. Behavior mapping data provided evidence that both user groups engaged in casual socializing with their neighbors in the hallways. Long-corridor residents' complaints about lack of privacy and inability to regulate their local social contacts were apparently a function of the number of people forced to share the central corridor as both a circulation area and an interaction zone. Interestingly, despite the fact that long-corridor residents came in contact with neighbors more frequently, they reported greater difficulty making friends on their floor. This pattern is a reversal of previous findings in homogeneous student populations. Other investigators (e.g., Festinger,

Schachter, & Back, 1950) have found a positive relationship between interaction frequency and friendship. It appeared that the environmental conditions under which neighbors regularly met was largely responsible for this phenomenon. The lack of suitable semiprivate space between individual units combined with a relatively large local residential group inhibited the development of smaller social groups that could have made local social life more manageable and satisfying.

It was clear that residents of long-corridor dormitories were reporting residential experiences that paralleled those found in Stony Brook corridor-style housing. However, to determine whether the group size variable had similar effects in nonresidential settings, we felt it necessary to replicate findings obtained in laboratory studies that manipulated the probability of interaction. Thus, we conducted a modified waiting-room experiment closely resembling that previously used with the Stony Brook sample. In this variation of the waiting-room procedure, subjects from long and short corridors waited either alone or with a confederate with whom they expected to interact, but all subjects waited twice as long. As before, affective responses and behavior of the two groups were very similar when waiting alone. However, while waiting with a potential interactant, long-corridor residents felt significantly less comfortable, sat farther from, and looked less at the confederate. These data reflected long-corridor residents' attempts to insulate themselves from social contact and thereby reduce the likelihood of interacting with our confederate stranger. Our Stony Brook data seemed to indicate that the avoidance strategies adopted by socially overloaded residents from corridor-style dormitories were ineffective in restoring positive mood and reducing reported stress. Yet, the fact that these subjects were in the experimental situation for only 5 minutes precluded examination of possible adaptation over longer periods of time. With this in mind, we administered mood questionnaires twice to the Trinity subjects. The results showed that reported stress levels among long-corridor students expecting to interact with the confederate did not dissipate over time.

Indices of Social Pathology

Taken together, the results of our studies have provided evidence that architectural factors not only mediate the quality of residential social life but engender interpersonal orientations and mood that become persistent and generalized modes of responding. Recently, however, we have become concerned with other potential consequences of residential density. We are confident that high social density is likely to lead to a relatively negative social experience, but we have found ourselves asking whether the phenomenon has more serious implications for the quality of life other than those already observed. Differential response to interaction probability and qualitative differences in residential social relations are suggestive but seem to be equivocal indicators of social pathology. However, extrapolations from these data have served to generate additional hypotheses about potential costs to individual functioning and threats to security.

Two studies were designed to assess the consequences of chronic loss of control over interactions. We hypothesized that if the inability to control or regulate social contact in Stony Brook corridor-style and Trinity long-corridor dormitories was sufficiently debilitating, these socially overloaded residents might manifest general motivational decrements and symptoms of learned helplessness (Seligman, 1975). Results from these experiments verified our hypothesis. When trying to solve geometric puzzles similar to those used by Glass and Singer (1972) to assess aftereffects of noise stress, subjects from Stony Brook corridor-style housing gave up more quickly and seemed to have a lower threshold for frustration than students living in suites (Davis, 1977). Residents from Trinity's long-corridor dormitories responded in an analogous manner to another experimental situation. Subjects who resided in long-corridors were found to be less likely to ask questions about an ambiguous experimental situation, and when they could not interact with a confederate in a bargaining game, they were more likely to choose withdrawal responses resulting in loss of points than were residents of short-corridors. Thus, we have found evidence suggesting that residential social conditions can lead to differential frustration tolerance and to different orientations regarding choice and control.

Other research that we are currently pursuing deals with the relationships among design, the presence or absence of semiprivate interaction loci, and criminal victimization. The impetus for this type of research was provided by the earlier work of Yancey (1972) and Newman (1972). Newman found that the absence of semiprivate interaction zones between individual dwelling units, among other factors, was positively associated with crime in public housing. Our analyses of residents' space use and local group development in the Stony Brook dormitories have allowed us to generate reasonably specific predictions about the incidence of criminal victimization in these residential environments. Because we are still in the initial phase of this research, we will report only preliminary findings. We hypothesized that since the central hallway in corridor-style dormitories remains public space to which access is largely uncontrolled by users, residents of this dormitory type would be at greater risk to criminal entry into the local environment and subsequent victimization. The semiprivate suite lounge in suite-style housing provides an important buffer zone between public hallways and private bedroom areas. Access to the suite lounge is under the control of users, and residents can easily monitor and control who has direct access to private areas. Preliminary analyses have provided evidence in support of this basic hypothesis; a considerably higher proportion of floors in corridor-style housing experienced criminal victimization above the median rate than floors in suite-style housing during a recent 2½-year period. Further analyses will be directed toward assessing the influence of residents' characteristics in combination with architectural features on crime rate.

The Final Test: Intervention

We hope that this discussion has illustrated how a rather long-range research program has evolved. As can be seen, we have not limited evaluation and experimentation to a single paradigm. The payoffs of a multimethod approach and multilevel analysis have been considerable. The consistent and predictable effects of residential group size and the manner in which neighbors are clustered around shared space have increased our confidence in the validity of our theorizing to the point where it now seems appropriate to consider possible intervention strategies. In fact, we probably embarked on this program of research because we felt that what we would learn could be applied in a constructive manner.

The goals of site intervention are twofold. First, we seek, on the one hand, to improve the quality of residential social experience. Second, the implementation of a specific intervention will provide the opportunity to test and validate further our hypotheses concerning the consequences of environment–behavior interface. We feel that our data-based analysis of this "fit" strongly suggests that, although a range of design interventions may be appropriate, in order to be effective they must focus on individual and social experience and enhance individuals' regulation of social contact. The efficacy of a particular design modification depends on the extent to which it achieves three goals: (1) the grouping of appropriately sized clusters of residents around shared amenities and spaces, (2) the provision of adequate semiprivate interaction spaces that will support comfortable neighboring and minimize behavioral interference from others in the local environment, and (3) the development of protective local groups. Since we are not dealing with a strictly deterministic system in which design variables necessarily effect social experience in a direct manner, we cannot be assured that a design that meets the first two objectives will inevitably lead to neighbor intimacy or friendship. However, we feel that architectural modifications that fulfill the first two conditions will, at the very least, enhance the environmental conditions under which neighbors regularly meet and not actively *impede* effective group development.

The site of one of our interventions will be Trinity College's long-corridor housing. We are fortunate to have an enlightened college housing office that not only is interested in the social outcomes of the intervention but will also provide material support for this endeavor. The basic problem that the intervention must address is how best to subdivide the long-corridor environment into smaller residential clusters. We are currently considering a number of ways in which this might be accomplished.

The simplest and least expensive interior modification is to divide each floor approximately in half by placing an unlocked swinging door midway in the central corridor. This would create two distinct residential areas with their own baths and access routes to the outside. Another intervention that has been considered involves conversion of two mid-corridor bedrooms into a combination lounge-kitchen to be shared by those living on the floor. Although each of these proposals is of merit, we feel that each addresses only a specific aspect of the problem and therefore, in our judgment, may not prove to be efficacious. The provision of an unlocked swinging door may be an effective facilitator of moderate-sized groups of residents and may reduce the probability of unwanted or unmanageable interactions with local others. This intervention, however, is visually intrusive and may not facilitate evolution of semiprivate zones with the requisite amenities to support casual socializing. The second proposal was designed to meet this objective but would do so at the expense of congestion and potential behavioral interference associated with a large number of people sharing the common lounge-kitchen areas.

Thus the most likely intervention is a combination of these proposals. By subdividing the floor and providing lounge areas to each cluster of residents, group size can be reduced and semiprivate interaction zones provided. If our analyses have been accurate, such an intervention should improve the quality of life in the long-corridor environment. Social avoidance and stress should be reduced in dormitory and laboratory settings.

As researchers, such an opportunity to test our ideas is rather frightening. Although success would be gratifying in terms of both theoretical corroboration and practical contribution, it could also lead to overly simplified treatments of residential density. Deterministic perspectives are tempting ones, but they necessarily describe and predict behavior on the basis of limited and often isolated variables. While our interventions in dormitories may be successful, it is important not to rush to judgment and apply indiscriminantly these findings to other settings. Because of the highly complex nature of environment–behavior relationships, studies of processes in one place may not be good predictors of outcomes in other places (e.g., Altman, 1973). Until this kind of analysis is successfully applied to other "places," conceptual linkages and design recommendations should not be routinely generalized. However, we feel that our experiences during our program of research suggest interesting ways of viewing environment–behavior relationships and of studying them. By systematically studying environmental influence at several levels of analysis, we may learn a great deal about how the social and physical environment mediates experience and behavior and we may be able to meaningfully influence the design of more appropriate settings.

ACKNOWLEDGMENTS

The authors would like to thank Thomas Smith (vice-president), David Winer (dean of students), and Kristina Dow (director of college residences), all at Trinity College, for their invaluable assistance and support.

References

Altman, I. Some perspectives on the study of man-environment phenomena. *Reports of Research in Social Psychology*, 1973, 4(1), 109–126.

Baum, A., & Koman, S. Differential response to anticipated crowding: Psychological effects of social and spatial density. *Journal of Personality and Social Psychology*, 1976, 34(3), 526–536.

Baum, A., Mapp, K., & Davis, G. E. Determinants of residential group development and social control. *Environmental Psychology and Nonverbal Behavior*, 1978, 3, 145–160.

Baum, A., & Valins, S. *Architecture and social behavior: Psychological studies of social density*. Hillsdale, New Jersey: Erlbaum, 1977.

Calhoun, J. Population density and social pathology. *Scientific American*, 1962, 206, 139–148.

Calhoun, J. Ecological factors in the development of behavioral anomalies. *Comparative Psychopathology*, 1967, 1–51.

Calhoun, J. Space and the strategy of life. *Ekistics*, 1970, 29, 425–437.

Christian, J. The pathology of overpopulation. *Military Medicine*, 1963, 128, 571–603.

Christian, J., Flyger, V., & Davis, D. Factors in the mass mortality of a herd of Sika deer *cervus nippon*. *Chesapeake Science*, 1960, 1, 79–95.

Davis, G. E. *Environmental antecedents of learned helplessness*. Unpublished manuscript, State University of New York at Stony Brook, 1977.

Desor, J. Toward a psychological theory of crowding. *Journal of Personality and Social Psychology*, 1972, 21, 79–83.

Festinger, L., Schachter, S., & Back, K. *Social pressures in informal groups*. New York: Harper, 1950.

Freedman, J., Klevansky, S., & Ehrlich, P. The effect of crowding on human task performance. *Journal of Applied Social Psychology*, 1971, *1*, 7–25.

Galle, O., Gove, W., & McPherson, J. Population density and pathology: What are the relationships for man? *Science*, 1972, *176*, 23–30.

Glass, D. C., & Singer, J. E. *Urban stress*. New York: Academic Press, 1972.

Griffitt, W., & Veitch, R. Hot and crowded: Influences of population density and temperature on interpersonal affective behavior. *Journal of Personality and Social Psychology*, 1971, *17*, 92–98.

McCarthy, D., & Saegert, S. Residential density, social overload, and social withdrawal. In J. R. Aiello & A. Baum, *Residential Crowding and Design*. New York: Plenum, 1979.

Milgram, S. The experience of living in cities. *Science*, 1970, *167*, 1461–1468.

Newman, O. *Defensible space*. New York: Macmillan, 1972.

Rodin, J., & Baum, A. Crowding and helplessness: Potential consequences of density and loss of control. In A. Baum & Y. M. Epstein (Eds.), *Human response to crowding*. Hillsdale, New Jersey: Erlbaum, 1978.

Seligman, M. E. P. *Helplessness*. San Francisco: Freeman, 1975.

Schmitt, R. C. Implications of density in Hong Kong. *Journal of the American Institute of Planners*, 1963, *24*, 210–217.

Schmitt, R. C. Density, health, and social organization. *Journal of the American Institute of Planners*, 1966, *32*, 38–40.

Stokols, D. A social psychological model of crowding phenomena. *Journal of the American Institute of Planners*, 1972, *38*, 72–84.

Wolfe, M. Room size, group size, and density: Behavior patterns in a children's psychiatric facility. *Environment and Behavior*, 1975, *7*, 199–224.

Winkel, G. *The methodological implications of ecological validity for the behavioral sciences*. Symposium paper presented at the American Psychological Association meetings, Washington, D.C., August 1976.

Yancey, W. Architecture, interaction, and social control: The case of a large-scale housing project. In J. Wohlwill & D. Carson (Eds.), *Environment and the social sciences*. Washington, D.C.: American Psychological Association, 1972.

12

Design Implications of Spatial Research

Gary W. Evans

Several recent reviews of human spatial research attest to the marked growth of this area of inquiry (Altman, 1975; Baum & Epstein, 1978; Edney, 1974; Evans & Howard, 1973; Evans & Stokols, 1976; Sundstrom, 1978). While this research has increased our understanding of the role of physical space in human behavior, little effort has focused upon the problems of applying this research to the design and planning process itself. The aim of this chapter is to derive some design implications from pertinent human spatial research and to suggest some future research strategies to help bridge the gap between proxemic research and design (see Altman, 1975, Chapter 11, for a thoughtful discussion of the etiology of this gap).

This discussion focuses on the microenvironment, i.e., the application of spatial research to the design of immediate, interior spaces. Stokols (1973), Fischer, Baldassare, and Ofshe (1975), and others have considered application to macro spaces. Furthermore, Baum and Valins (1977) have focused upon more intermediate scale design issues at the multiple dwelling unit level exploring the differential impact of corridor as opposed to suite-design dormitories on students' perceptions of crowding and group cohesiveness.

Three general facts that have emerged from the spatial literature are relevant here. First, the human organism has certain spatial needs which

Gary W. Evans • Program in Social Ecology, University of California, Irvine, California 92717.

if not provided for can lead to stress. Second, there are considerable individual differences in human needs and preferences for physical space. Third, various psychological variables intervene between the physical parameter, space, and the behavioral responses of the individual. All three of these facts provide the designer little if any specific direction because, in sum, they simply suggest the principle that the best thing the designer can do is to provide variety and flexibility in spaces designed for human use. One hopes that most researchers in spatial behavior are aware that designers knew these principles a long time ago! What can one do to move beyond the litany of variety and flexibility? I would like to suggest two general areas of available proxemic data that may be helpful to the designer.

First, data are available that specify the conditions under which unmet spatial needs are more likely to occur. Put another way, some insights have been made into those individual characteristics, interpersonal situations, and social settings wherein more or less physical space is needed for comfort. These psychological variables will be only briefly discussed because the designer's major focus is on the manipulation of physical parameters. Psychological variables are critical, however, in that they provide some general guidelines for the context in which specific design decisions are made. For example, if one's task is to design a space for a certain group of users within a particular organizational framework, it is helpful to acquire existing data on certain personalogical and social setting variables which influence spatial needs. The purpose of the discussion on psychological variables is to suggest which psychological variables the designer might want to find out about before designing a space.

Second, there are a handful of empirical studies which have manipulated design parameters and measured the impact on perceived spatial impingements of personal space and crowding. These findings are discussed in detail. Before reviewing human spatial behavior relevant to design, an overview of the dominant theoretical perspectives is presented.

Brief Synopsis of Theoretical Perspectives in Human Spatial Behavior

An understanding of basic theoretical perspectives in spatial behavior is helpful in organizing the numerous variables that affect human spatial needs. Briefly, four perspectives have evolved: behavioral constraint, overload, ecological psychology, and ethological models. The *behavioral constraint* perspective emphasizes the importance of personal control in spatial behavior. An individual who perceives that his/her goals are thwarted by inadequate space is more likely to feel spatial impingement. One extension of this perspective is that perception of less control over the social or physical setting is likely to heighten individual needs for space or increase stress when inadequate space is available.

The *overload* perspective suggests that the close proximity of others is one form of information overload wherein the individual is bombarded with too much social and/or physical stimulation. Given the limited capacity of human beings to deal with information, at a certain point the amount of information which the organism is required to deal with exceeds that capacity and a state of overload results. The impact of overload on human social behavior and cognitive performance has been reviewed by Fischer (1976) and Cohen (1978). Cohen argues that in a state of overload, individuals cannot process as much information because of reduced attentional capacities. Among the consequences of this reduction in capacity is less sensitivity to peripheral information. This may lead to poorer performance, particularly on a difficult task. Fischer critically reviews urban determinism theory, which argues that urbanites adapt to the overload city environment by becoming socially withdrawn, showing less concern for others, and adopting a generally, cool, brusque interpersonal style.

The *ecological psychology* model emphasizes the ratio of the number of persons in a setting to the number of roles necessary to maintain that setting. When a setting becomes overpersonned, there are more people available than necessary to maintain operations. An important extension of the ecological concept stresses the importance of resource scarcity to feelings of crowding. When the ratio of numbers of persons to numbers of resources grows too great, negative feelings may occur.

Finally, the *ethological* model has stressed the link between spatial needs and human adaptation. This view of human spatial behavior emphasizes the importance of space in terms of threats to individual autonomy and aggression. Impingements of space are seen as stressful because they may lead to fear and discomfort because of increased feelings of aggression and/or threat.

The critical link among these four theoretical perspectives is the *stress* construct. The physical variable space, plus the intervening psychological constructs of personal control, information capacities, roles, and concern about threat may interact to produce stress in humans. Again, a basic truism in human spatial behavior is that one can maximize their understanding of the likelihood of stress from insufficient space if one considers both changes in available physical space and variations in certain relevant psychological constructs (Evans & Stokols, 1976).

Adequate Space Provision: Behavioral Variables

This section examines personal space and crowding research on individual characteristics, interpersonal situations, and social settings. Primarily secondary sources are cited because of the large amount of data summarized.

Individual Characteristics

Personality Variables. Several personal space studies have examined personality variables. The following variables have been associated with smaller zones of personal space: internal locus of control (especially in presence of strangers), extroversion, less anxiety, positive self-concept, greater affiliation needs, and greater change and variety seeking (Altman, 1975; Evans & Howard, 1973). Density data indicate a greater tolerance for crowding among persons with internal loci of control (Sundstrom, 1978).

Abnormal Personality. The personal space research on abnormality is complex, with mild support for the position that those with greater abnormality have greater spatial needs (Evans & Howard, 1973). Data also indicate that more violent criminal offenders tend to have greater personal space zones as well. Nevertheless, there are sufficient contradictory findings indicating abnormally close encounters by mental patients to lead to the preliminary conclusion that the personal space behavior of abnormals is more variable than that of normals (Altman, 1975). More consistently, data have indicated that when one interacts with another individual labeled as physically or mentally abnormal, normal individuals choose greater interaction distances (Altman, 1975; Evans & Howard, 1973). Few or no data are available on crowded abnormal samples.

Sex. As Altman (1975) has pointed out, sex variables in spatial research have usually been treated as secondary in various analyses of spatial behavior, often being examined in combination with age, racial, or cultural variables. Overall, crowding data are equivocal, with many studies finding no sex differences (Sundstrom, 1978). Personal space research generally indicates that adult men require greater space than do adult women (Evans & Howard, 1973). Recent data suggest that the direction of spatial invasion may be critical. Women seem less concerned about frontal spatial intrusion as opposed to lateral spatial invasions. The opposite holds true for men (Fisher & Byrne, 1975).

Age. The overall trend in age data suggests that younger children are more susceptible to crowding than are adults (Evans, 1978a). Personal space research indicates very mixed developmental findings. Most *in vivo* studies have found that younger children interact more closely than older children and that adult spatial norms are adopted around puberty (e.g., Aiello & Aiello, 1974; Tennis & Dabbs, 1975). Nevertheless, several projective studies, e.g., placing silhouette cutouts on a board, have found decreases in space as children age (Evans & Howard, 1973). Given the greater validity of the *in vivo* techniques, I would offer the preliminary conclusion of increasing interaction distances as children grow up.

Cross-Cultural.   As the reader of Hall (1966) knows, there has been extensive discussion of the importance of cultural norms in human spatial behavior. Empirical studies have yielded moderate support of Hall's observations that "contact cultures" interact more closely than more distant North Americans and Scandinavians (Aiello & Thompson, in press; Altman, 1975; Evans & Howard, 1973). Some speculations on how certain cultures handle high density (e.g., Japan, Hong Kong) have ensued. Some have suggested that cultural differences in handling crowding derive from established privacy norms which rely upon highly regulated interaction patterns coupled with social hierarchies (Altman, 1975).

Past Experience.   Persons who have backgrounds of intensive, frequent social interaction appear more tolerant of crowding (Sundstrom, 1978). On the other hand, Cozby (1973) reported that persons of urban versus rural background exhibited no differences in personal space requirements but that individuals who grew up in higher density households did have larger personal space zones. Recent data suggest that persons who have larger personal space zones are less tolerant of crowding (Sundstrom, 1978). Furthermore, an immediately previous exposure to crowding has been shown to increase social withdrawal and create greater personal space needs in short-term laboratory conditions (Sundstrom, 1978).

Interpersonal Situations

People who are friends or view one another positively tend to interact more closely. On the other hand, extreme closeness can be used to threaten another. Data generally indicate that external threats, fear, or stress tend to increase an individual's need for space.

Crowding researchers have found that feelings of crowding tend to increase under the following conditions: solitary as opposed to social activities, evaluative climate, presence of larger numbers of persons, and sometimes being interfered with while working on a task (Sundstrom, 1978). In addition, crowding can be intensified by invasions of personal space.

Social Setting

A few studies have examined setting variables of a more social nature. Interpersonal distancing is greater in more formal settings and when working on less pleasant tasks. Further, when less structure is anticipated, subjects tend to feel more crowded. One's experience and investment in a setting that may be related to territorial-like experiences are related to spatial behavior. Edney (1972) concluded that subjects who had previous exposure to a room or who anticipated future use of a room were more likely to invade a stranger's space more closely in that room in order to express territorial threat. Castell (1970) found that children stood much closer to their mothers in a novel versus a familiar setting. These two bits of data are consistent with an important analysis by Stokols (1976) of primary and secondary environments in the experience of crowding. According to his analysis, crowding is likely to be more severe in primary environments than in secondary ones. Primary environments are those spaces in which the individual has more psychological investment, spends most of her/his time, and feels control over. Crowding is said to be more severe in a primary space because a greater range of potential security threats exists, as does the greater probability of loss of control over personal needs and goals.

Summary of Behavioral Variables and Spatial Needs

The designer is urged to consider behavioral variables that can affect human spatial needs. The above overview cannot replace a more thorough reading of the literature but may raise the saliency of variables that one must consider when determining spatial needs. This information may guide designers in deciding when more space is probably going to be needed as opposed to operating simply within minimum spatial regulations. Although the most direct design response to greater spatial needs is to increase the amount of space available, this strategy is not always feasible. For some time designers have been familiar with techniques which increase perceived space. Recent empirical research has explored design parameters which may ameliorate the effects of spatial restriction.

Adequate Space Provision: Design Variables

Room Dimensions

Although it may seem intuitively obvious, it is important to demonstrate that reductions in room size do, in fact, create feelings of increased spatial restriction. Smaller room size can cause crowding (Sundstrom, 1978), although some research has found slight or no effects of reduced room size on some behaviors (Freedman, 1975; see Evans & Eichelman, 1976, for a discussion of this controversy in the literature). Both room size and degree of enclosure also affect personal space behavior. Savinar (1975) reduced ceiling height and found that males had concomitantly greater need for space when approached by an experimenter. Daves and Swaffer (1971) determined the distance at which subjects would move away from an approaching experimenter in a large (12 feet × 29 feet), a small (4 feet × 6 feet), and a long and wide space (7.5 feet × 65 feet), tested at both wide and narrow wall. They found the greatest distances in the large and long spaces followed by the wide and small spaces, respectively. Desor (1972), in a projective crowding study, found that when people are instructed to place stick figures in an interior scale model up to the point at which the room would become crowded, they placed more figures in rectangular models than square ones with area held constant. More recently White (1975) reported significant increases in personal space with reductions in room size. The variance of personal space behavior was also significantly affected by room size in that greater variability was found in large rooms. Further inspection of the data indicated a sizable minority of persons whose personal space was larger in the large room, thus creating a somewhat bimodal distribution of scores. Nevertheless, overall subjects interacted more closely in the larger room.

Enclosure has been examined by measuring the personal space zones of persons while interacting in the center versus the corner of a room. Dabbs, Fuller, and Carr (1973) found that both college students and prison inmates had larger personal space zones in the corner of a room as opposed to the center. Tennis and Dabbs (1975) replicated the corner–center difference across an age span of 5th-year, 9th-year, 12th-year, and college sophomore-year students, with a reversal found for first graders. Finally, Brody and Zimmerman (1975) found that students from open classrooms in comparison to others from traditional classrooms tended to have smaller personal space zones, as indicated by closer silhouette placement of figures representing acquaintance, bully, and visitor, with no differences found for best friend or teacher. It is conceivable that exposure to the less enclosed space of the open classroom may have affected spatial behavior. Other plausible explanations are apparent as well, including differences in educational philosophies expressed by teachers or actual interaction distances employed by teachers in the respective classrooms.

One's perception of openness in a space may be affected by many factors other than the actual physical enclosure. Garling (1969) reported that perceived openness increases with more physical size *in situ* and in photographs. Hayward and Franklin (1974), however, found that the ratio of boundary wall height to wall distance is a more critical architectural factor in enclosure perception. It would be interesting to extend these findings to the study of proxemic behaviors. For example, would an individual exhibit greater need for space when perceived enclosure was reduced even though physical enclosure was held constant?

Indoor and Outdoor Space

A sense of enclosure is one factor which presumably varies between indoor and outdoor spaces. A few studies have looked at the relative impact of indoor and outdoor spaces on spatial behavior. Little (1965) found that subjects tended to project smaller interaction distances in an open-air setting than in indoor settings. Similarly, Pempus, Sawaya, and Cooper (1975) reported that subjects *in vivo* tended to keep a greater distance between themselves and a stranger in an indoor setting than in an outdoor space.

Dabbs *et al.* (1973) suggested that perhaps people need more space with enclosure because they perceive greater threat from the marked reduction in possible escape routes available. McClelland and Auslander (1976) reported that subjects rated scenes which contained more visual escapes (windows and doors) as less crowded. Addressing this issue more indirectly, Schiffenbauer, Brown, Perry, Shulack, and Zanzola (undated) hypothesized that dormitory residents on higher floors might feel less crowded (physical space held constant) because of greater visual escapes, i.e., a better view. Their data indicated that while those on higher floors perceived that they had larger rooms, they did not feel less crowded than those on lower floors. On the other hand, the feeling of being trapped on a higher floor could counterbalance the feeling of a larger room because of greater vistas. The visual escape explanation is not the only plausible explanation of the relationship between visual escape and spatial needs. A better view, the presence of more doors and windows, etc., also provide greater distraction or visual complexity in the setting. Recently several investigators have explored complexity and distraction issues in the spatial literature.

Visual Complexity and Distraction

Coss (1973) has argued that the use of deliberate distractions in design may be helpful in the reduction of stress in high arousal-producing settings. Considerable evidence indicates that restrictions of space often lead to overarousal in humans that is experienced as stressful (Evans, 1978b). The placement of objects on walls, windows with views, advertisements in mass transportation vehicles, fireplaces, and the availability of small objects, games, etc., to manipulate are some of the displacement stimuli suggested by Coss. The basic idea is that these mechanisms act as "cutoff" devices which provide individuals with something to distract their attention from the more arousal-producing elements of the situation.

Baum and Davis (1976) employed a projective modeling technique and found that the placement of pictures on a wall tended to decrease crowding only in a social situation (party) as opposed to a less social setting (airport). This finding was further qualified by an interaction with wall brightness (light vs. dark green). Subjects placed significantly more figures in model dark rooms when the activities were social. Subjects also placed more figures in the light room under social conditions, but the difference was not significant. Further, when the space was not social, the combination of dark colors plus visual complexity served to increase crowding intensity. Worchel and Teddlie (1976) also manipulated the presence of wall pictures except in a live setting. They found that the presence of pictures was particularly salient in reducing discomfort when personal space was invaded. The addition of pictures reduced experiences of crowding, confinement, and tension in close, but not far, interaction distance conditions.

These findings on visual complexity or distraction must be treated with caution for two reasons. First, in each of these studies, there were no main effects of the distraction dimension, i.e., complexity interacted with other factors to sometimes produce a more positive effect when space was limited. At this time the robustness of the complexity factor must be considered limited. Second, considerable perceptual research and theory indicate that persons typically prefer a moderate amount of environmental or stimulus complexity (Berlyne, 1960). This suggests that we must be careful in *how much distraction* we attempt to provide in a setting so as to reduce the stress of overcrowding or spatial invasion. Presumably, there is an optimum level of distraction beyond which the presence of too many cutoff stimuli could increase our discomfort.

The complexity-distraction data provide an interesting theoretical dilemma. In light of the overload perspective on human spatial behavior one might expect spatial restriction, plus more distraction, to increase overall stress because of the greater information levels present. Alternatively, the cutoff hypothesis suggests that the threat is reduced, since one is less aware of the close presence of others or of their spatial restrictions when attention is diverted away from the "other."

Brightness

Both Baum and Davis (1976) and Schiffenbauer *et al*. (undated) have found that light color or well-lit rooms, respectively, were perceived as less crowded than comparable, darker rooms. Carr and Dabbs (1974) have also reported that reducing room lighting may cause strangers to feel more uncomfortable. Subjects waited longer to answer an interviewer's questions and made shorter visual glances at the interviewer in dim as opposed to bright light.

Room Partitioning

The overload perspective on spatial behavior suggests that the presence of partitions might reduce stress from spatial restriction since partitions would help cut down on visual exposure, noise, and other sources of stimulation. Partial support for this position derives from findings by Desor (1972) and Baum, Riess, and O'Hara (1974), who have demonstrated that partitions reduce crowding and feelings of spatial invasion, respectively. Desor reported that people placed more stick figures in scale-model rooms when partitions were present. Curiously, it made no difference whether the partitions were transparent or opaque, full or half height. Baum *et al*. (1974) found that subjects would drink from a water fountain significantly less often when a confederate was within five feet of the fountain. The presence of a screen around the fountain, however, significantly increased the probability that someone would drink from the fountain when the confederate was nearby.

Stokols, Smith, and Proster (1975), however, demonstrated that partitions in a crowded waiting area slightly increased feelings of crowdedness and significantly increased behavioral indices of tension. They suggested that individuals may have viewed the partitions as herding devices which restricted their behavioral options in the setting and thus could have led to greater discomfort.

At this time, it seems cogent to carefully weigh the potential psychological and physical impact of partitions on the individual in a spatially restricted setting. In situations wherein the individual does not desire or is not concerned with behavioral control, partitions may help reduce perceived spatial restriction or at least its discomforts. Perhaps allowing users of a setting to manipulate the position of partitions could maximize the protective shielding effects of partitions while preserving a sense of behavioral options. Many writers have noted the apparent successful use of movable partitions by the Japanese to regulate privacy in situations in which little space is available.

Summary of Design Variables and Spatial Needs

Several design interventions have been discussed which may modify the individual's reactions to inadequate space. The size and shape of rooms, their degree of openness, access levels to the external environment, complexity and displacement, brightness, and extent of partitioning are the design variables that have had some preliminary exploration.

While one familiar with the spatial literature cannot help but be encouraged by the recent increase in proxemic studies that explore design variables, the potential user of these data might well be perplexed by the complexity of the findings and concerned about the extent of their reliability and validity in real design settings. Some of the design and proxemic data are equivocal, if not contradictory, and may be limited in their potential application because of their laboratory origins. In the next section some of these problems are discussed and some alternative research strategies are proposed.

Proxemics and Design: Previous Faults and Some Future Ideas

The most apparent limitation in the research discussed in this chapter is that nearly all of it could be classified into behavioral or physical variables. Both sets of studies support the claim that both physical and psychological variables are important in human spatial behavior. Therefore, it seems cogent to explore how these variables operate together.

A second deficiency in the proxemic and design research has been the limited time parameters employed. With a few exceptions, we are implicitly asking designers to make design modifications which have a rather permanent quality on the basis of data derived from very short passages of time. Several of the theoretical perspectives on spatial behavior discussed earlier emphasize the importance of temporal variables. The linking construct of stress in particular is pertinent. Acute versus chronic stressors and the monitoring of concurrent versus delayed effects are but two important variables in stress research that also are relevant in the human spatial literature (Evans & Eichelman, 1976). The complexity of the human response to stressors such as crowding or personal space invasion further demands multivariate analysis techniques. We need not only to explore subjective self-reports (e.g., semantic differentials) but also to monitor behavioral responses (e.g., simple and complex task performance), unobtrusive observational indices (e.g., behavioral stereotypes), and psychophysiological changes (e.g., skin conductance).

The need for more time sampling in human spatial research is matched by problems of setting specificity in our research. As Brunswik urged psychologists some time ago, it is important for behavioral scientists not only to select representative samples of people but also to carefully select representative samples of settings (Brunswik, 1949). The relevance of this point to the discussion here is illustrated by the contradictory findings of Stokols *et al.* (1975) regarding the use of partitions to reduce crowding in high density settings. Laboratory simulation studies had indicated that partitions tended to reduce crowding stress, whereas Stokols's data indicated the opposite. This contradiction does not simply suggest that field settings may not be the same as laboratory manipulations. First, the lab studies with partitions were projective tests (putting stick figures in models). Second, the setting employed by Stokols *et al.* (1975) was a public waiting area where people lined up to obtain services from a large, somewhat impersonal bureaucratic organization. The negative impact of partitions in that setting might not generalize to other field settings where people are not waiting in lines for services.

The setting issue is not one of the oft-repeated, highly simplistic laboratory-versus-real-world comparisons. Rather, it is important to understand which setting dimensions interact with particular personal characteristics and needs. Both physical and social aspects of settings can interact with personal characteristics to affect behavior. Conceivably, certain laboratory-versus-real-world differences are unimportant for certain behavioral outcomes. Furthermore, laboratory and real world settings may not differ at all on certain dimensions or differ to a lesser extent than other modes of setting definition and classification. The task then becomes one of identifying the critical setting and personal characteristics which contribute to particular behaviors of interest.

The theoretical perspectives in human spatial behavior briefly outlined earlier provide some additional preliminary thoughts for continuing research on proxemics and design. The application of the behavioral constraint or control approach is one critical component of Stokols's (1976) recent formulation of the primary–secondary environment distinction. The reduction of personal control vis à vis spatial restriction may be most critical in primary settings precisely because one expects and is accustomed to maximum control in such settings. The primary–secondary distinction also lends support to the importance of temporal factors in proxemic and design research, since primary environments are defined in part on the basis of how long an individual has been associated with a given setting as well as how long he/she typically spends in that setting. Designers must examine the degree of occupants' overall perceived control in a setting as well as consider design solution which can accommodate user needs for the exercise of behavioral control over space.

The overload perspective leads directly to the examination of partitions in high density environments as one means to reduce overload. The possible trade-off between reducing overload and increasing perceived constraint from partitions has complicated this proposed design modification in high density situations.

Let us examine a particular design problem to illustrate how the behavioral control perspective and the overload model might be applied. The problem is to design a dormitory room for two people. (See Van Der Ryan & Silverstein, 1972, for a sensitive discussion of user evaluations of dormitories.) If we had generous funding, one potentially optimal solution would be to provide each person with a private bedroom space, each of which opened onto a common study-lounge space that the two roommates would share. This design would provide highly personal spaces where high personal control and maximum privacy are easily maintained. At the same time, this design also provides a shared social space where interaction and general stimulation levels could be increased.

Typically, such a design solution is unlikely because of budget restraints and one is left with the difficult situation of one room to be shared by two persons. A frequent problem with this arrangement is the lack of privacy it affords roommates. Two possible options which might improve the situation are as follows:

First, if one is not saddled with a square or rectangular constraint (i.e., four flush walls), one option is to provide a space shaped like two L's that are back to back and in opposite orientation. Each small alcove (the bottom of the L) could have either a sliding door or some type of movable partition. This option still retains some of the advantages of the former design solution with the provision of at least a minimum amount of highly personal space. Even if this personal space were exceedingly small, I suspect it would be preferable to a larger room with no such personal space and only a larger, common space where one had to share control with one's roommate.

A final example is one in which the designer must work with four flush walls. Here one could institute a modifiable partition system in which roommates could choose from a selection of different sized and shaped partitions that could be attached from floor to ceiling at most points in the dorm room. To assist students in making their choices, photographs (drawings) and descriptions of various arrangements could be provided. One critical aspect of this option is the modifiability of the design wherein roommates can rearrange their environment as interpersonal needs change. Furthermore, building codes permitting, qualitative differences in partitions could be available also. Partitions of different height, color, and material, and with windows or doors, etc., could be utilized.

Another design response to the overload position might be to examine the approach of Berlyne (1960) and recent extensions by Mehrabian and Russell (1974), who have enunciated various physical dimensions that provide more or less information to the user. Theoretically, if crowding or personal space invasion increases stimulus input and thus contributes to overload, one might be able to cut down on overload by providing environments low in information content.[1]

The ecological perspective points to the importance of resource availability and roles in settings. One might reduce feelings of crowding or insufficient space in a dormitory setting by making sure that students have enough study, lounge, etc., resources at hand. Programmatically, the increased sense of belonging and self-importance emanating from participating in setting maintenance suggests that designers should consider options which will increase the number of roles required to maintain a setting. One way is to set firm limits on the upper bounds of social units with which participants will most likely identify. Perhaps physically separating and demarcating subsections of large dormitories, plus structuring social and governance units in accord with those subareas might increase a sense of the individual's participation and importance in the maintenance of the setting. Conceivably, one reason why Baum and Valins (1977) have consistently found corridor-style dorm inhabitants less enamored of their living arrangements than those who live in suites is that more individual recognition and self-importance is perceived in the suite situation.

Finally, the importance of threat and aggression in a social context may suggest design alternatives which can ameliorate the stresses of inadequate spatial resources. First, security provision may diminish perceived threat. Second, given the arousal-producing aspects of threat and/or aggression, the use of deliberate arousal-reducing mechanisms (Coss, 1973) may help. Zimring, Evans, and Zube (1978) have discussed

---

[1]Overload is considerably more complex than is presented here or in most other works. In addition to quantitative shifts in information load, one must also consider qualitative differences in overload. For example, certain stimuli are more demanding than others, which suggests certain information-processing biases in human beings (Kaplan, 1973). Furthermore, although most overload theorists have focused on *sensory* overload, the importance of *social* overload has not been clearly delineated. See Cohen (1978), Evans and Eichelman (1976), and Fischer (1976), for more detailed discussions of overload theory.

in some detail the utility of the arousal construct as a theoretical link between proxemics and design with an extended discussion of arousal-reducing design mechanisms.

In dormitories, perceived security could be enhanced by providing both greater physical security and more easily monitored transition spaces between building entrances and living areas (Newman, 1974). For example, if an intruder had to pass through a lounge or study area before reaching individuals' rooms, there might be more deterrence. This strategy would probably be particularly effective if coupled with the ecological psychology perspective wherein relatively small, clearly identified social units subdivided from the dormitory coincide with particular physical areas of dormitory space. Thus, rather than have a large ground-floor lounge that is undefined in terms of ownership and where few, if any, people are known to all as part of the dorm, smaller lounges should be dispersed throughout the dorm proximate to physically and socially identifiable dormitory subsections.

For the most part, this discussion of proxemics and design has neglected the concept of territory. The relationship between territorial dynamics and the protection and defense of spaces at a macrolevel has been elaborated on by Newman (1974). One important aspect of territorial behavior which may be of relevance to more microspaces is the importance to the territorial holder of a clearly marked and bounded space that can readily be identified as his/her space, as opposed to someone else's space. Human beings engage in various forms of marking behavior which suggest both that individuals respect and react to other's markers and that markers are used to indicate a sense of ownership or control over a given space (Altman, 1975). The importance of clearly discernible boundaries in our cognitive representations of larger areas has also been stressed (Kaplan, 1973).

Two suggestions for design follow: (1) It is important to supply hetereogenous spaces where possible so that individuals can readily identify and recognize different spaces; (2) programmatically, it also becomes important to provide individuals the freedom to personalize their immediate, primary environments. Sommer (1974) has pointed out how important it is for individuals to be able to personalize their living spaces to gain a sense of belonging and identity with a space.

To continue our dormitory illustration, one way in which smaller subunits of the dormitory could be enhanced is to provide both social and physical support structures that reinforce external distinctiveness and internal cohesion and identification. Simple things, such as dormitory governmental structures or common-interest subsections of dormitories, are examples of social structures that might facilitate distinct subcultures in a dormitory. Physically, one could provide different colors, textures, furniture, motifs, etc., for different areas. Clear boundaries marked with doors, signs, or obvious transition zones could also reinforce designated social substructures in the dormitory.

Finally, dormitory policies that facilitate room personalization are to be encouraged. Allowing students to decorate their rooms is a minimum requirement. Greater sense of personalization and personal control might also follow from the use of the modifiable partition system suggested above. In addition, the provision of room furniture banks where students can select room furniture on a loan basis could increase personalization.

All the various examples of dormitory design and policy options are obviously preliminary suggestions which are speculative. Nevertheless, they illustrate options that are sensitive to the various emerging theoretical perspectives on spatial behavior discussed in this chapter.

Clearly our understanding of human spatial behavior is not yet at a stage in which one can conceptually formulate or empirically demonstrate integrated design programs that complement human spatial needs. Nevertheless, the recent increase in proxemic studies which have examined design-related variables is a particularly hopeful trend that may lead proxemic researchers beyond the well-meant but essentially noninformative calls for more variety and flexibility in design.

Summary

The purpose of this chapter is to provide designers with available information on human spatial needs and to outline some design strategies that can speak to those needs. Both psychological and physical variables which impact spatial behavior are reviewed. Limitations in this research and some suggestions for future work are presented. The importance of simultaneously examining changes in psychological and physical variables as they affect spatial behavior over varying periods of time and in a variety of settings is emphasized.

One way in which the issues of limited temporal and setting sampling can be addressed is by the adoption of an alternative research process. Perhaps researchers and designers could collaborate during the design formulation process. Theoretically derived or preliminary lab data-based hypotheses could be tested by systematically injecting them as design and program variables in existing or proposed settings. This would provide the opportunity for prospective studies, plus the normal opportunities for postconstruction user evaluations.

ACKNOWLEDGMENTS

In addition to the editors of this book, I thank Sheldon Cohen, Allen Schiffenbauer, Daniel Stokols, Craig Zimring, and the environmental psychology group at the University of Massachusetts at Amherst for their comments on earlier drafts.

References

Aiello, J. R., & Aiello, T. D. The development of personal space: Proxemic behavior of children 6 through 16. *Human Ecology*, 1974, 2, 177–189.

Aiello, J. R., & Thompson, D. E. Personal space, crowding, and spatial behavior in a cultural context. In I. Altman, J. F. Wohlwill, & A. Rapoport (Eds.), *Human behavior and environment* (Vol. 4), *Culture and environment*. New York: Plenum Press, in press.

Altman, I. *The environment and social behavior: Privacy, territoriality, crowding and personal space*. Monterey, California: Brooks/Cole, 1975.

Baum, A., & Davis, G. E. Spatial and social aspects of crowding perception. *Environment and Behavior*, 1976, 8, 527–544.

Baum, A., & Epstein, Y. (Eds.). *Human response to crowding*. Hillsdale, New Jersey: Erlbaum, 1978.

Baum, A., Riess, M., & O'Hara, J. Architectural variants of reaction to spatial invasion. *Environment and Behavior*, 1974, 6, 91–100.

Baum, A., & Valins, S. *Architecture and social behavior: Psychological studies of social density*. Hillsdale, New Jersey: Erlbaum, 1977.

Berlyne, D. E. *Conflict, curiosity and arousal*. New York: McGraw-Hill, 1960.

Brody, G. H., & Zimmerman, B. J. The effects of modeling and classroom organization on the personal space of third grade and fourth grade children. *American Educational Research Journal*, 1975, 12, 157–168.

Brunswik, E. *Systematic and representative design of psychology experiments*. Berkeley: University of California Press, 1949.

Carr, S., & Dabbs, J. The effects of lighting, distance, and intimacy of topic on verbal and visual behavior. *Sociometry*, 1974, 37, 592–600.

Castell, R. Effect of familiar and unfamiliar environments on proximity behavior of young children. *Journal of Experimental Child Psychology*, 1970, 9, 342–347.

Cohen, S. Environmental load and the allocation of attention. In A Baum & S. Valins (Eds.), *Advances in environmental psychology*. Hillsdale, New Jersey: Erlbaum, 1978.

Coss, R. The cut off hypothesis: Its relevance to the design of public places. *Man-Environment Systems*, 1973, 3, 417–440.

Cozby, P. Effects of density, activity and personality on environmental preferences. *Journal of Research in Personality*, 1973, 7, 45–60.

Dabbs, J., Fuller, P., & Carr, S. *Personal space when cornered: College students and prison inmates*. Presented at the meeting of the American Psychological Association, Montreal, 1973.

Daves, W., & Swaffer, P. W. Effects of room size on critical interpersonal distance. *Perceptual and Motor Skills*, 1971, 33, 926.

Desor, J. Toward a psychological theory of crowding. *Journal of Personality and Social Psychology*, 1972, 21, 79–83.

Edney, J. J. Place and space: The effects of experience with a physical locale. *Journal of Experimental Social Psychology*, 1972, *8*, 124–135.

Edney, J. J. Human territoriality. *Psychological Bulletin*, 1974, *81*, 959–975.

Evans, G. W. Crowding and the developmental process. In A. Baum & Y. Epstein (Eds.), *Human response to crowding*. Hillsdale, New Jersey: Erlbaum, 1978. (a)

Evans, G. W. Human spatial behavior: The arousal model. In A. Baum & Y. Epstein (Eds.), *Human response to crowding*. Hillsdale, New Jersey: Erlbaum, 1978. (b)

Evans, G. W., & Eichelman, W. Preliminary models of conceptual linkages among some proxemic variables. *Environment and Behavior*, 1976, *8*, 87–116.

Evans, G. W., & Howard, R. B. Personal space. *Psychological Bulletin*, 1973, *80*, 334–344.

Evans, G. W., & Stokols, D. (Eds.). Theoretical and empirical issues with regards to privacy, territoriality, personal space, and crowding. *Environment and Behavior*, 1976, *8*(special issue).

Fischer, C. *The urban experience*. New York: Harcourt, 1976.

Fischer, C., Baldassare, M., & Ofshe, R. Crowding studies and urban life: A critical review. *Journal of the American Institute of Planners*, 1975, *31*, 406–418.

Fisher, J. D., & Byrne, D. Too close for comfort: Sex differences in two invasions of personal space. *Journal of Personality and Social Psychology*, 1975, *32*, 15–21.

Freedman, J. *Crowding and behavior*. San Francisco: Freeman, 1975.

Garling, T. Studies in visual perception of architectural spaces and rooms. II. Judgments of open and enclosed space by category rating and magnitude estimation. *Scandinavian Journal of Psychology*, 1969, *10*, 257–268.

Hall, E. T. *The hidden dimension*. New York: Doubleday, 1966.

Hayward, S. C., & Franklin, S. S. Perceived openness-enclosure of architectural space. *Environment and Behavior*, 1974, *6*, 37–52.

Kaplan, S. Cognitive maps in perception and thought. In R. Downs & D. Stea (Eds.), *Image and environment*. Chicago: Aldine, 1973.

Little, K. Personal space. *Journal of Experimental Social Psychology*, 1965, *1*, 237–247.

McClelland, L., & Auslander, N. *Determinants of perceived crowding and pleasantness in public settings*. Paper presented at the seventh annual Environmental Design Research Association, Vancouver, 1976.

Mehrabian, A., & Russell, J. *An approach to environmental psychology*. Cambridge, Massachusetts: M.I.T. Press, 1974.

Newman, O. *Defensible space*. New York: Macmillan, 1974.

Pempus, E., Sawaya, C., & Cooper, R. E. *"Don't fence me in": Personal space depends on architectural enclosure*. Paper presented at the 83rd Annual Convention of the American Psychological Association, Chicago, 1975.

Savinar, J. The effect of ceiling height on personal space. *Man-Environment Systems*, 1975, *5*, 321–324.

Schiffenbauer, A. I., Brown, J. E., Perry, P. L., Shulack, L. K., & Zanzola, A. M. *The relationship between density and crowding: Some architectural modifiers*. Unpublished manuscript, Virginia Polytechnical Institute, undated.

Sommer, R. Looking back at personal space. In J. Lang, C. Burnette, W. Moleski, & D. Vachon (Eds.), *Designing for human behavior: Architecture and the behavioral sciences*. Stroudsburg, Pennsylvania: Dowden, Hutchinson & Ross, 1974.

Stokols, D. The relation between micro and macro crowding phenomena: Some implications for environmental research and design. *Man-Environment Systems*, 1973, *3*, 139–149.

Stokols, D. The experience of crowding in primary and secondary environments. *Environment and Behavior*, 1976, *8*, 49–86.

Stokols, D., Smith, T., & Proster, J. Partitioning and perceived crowding in a public space. *American Behavioral Scientist*, 1975, *18*, 792–814.

Sundstrom, E. Crowding as a sequential process: Review of research on the effects of population density on humans. In A. Baum & Y. Epstein (Eds.), *Human response to crowding*. Hillsdale, New Jersey: Erlbaum, 1978.

Tennis, G. H., & Dabbs, J. M. Sex, setting and personal space: First grade through college. *Sociometry*, 1975, *38*, 385–394.

Van Der Ryan, S., & Silverstein, M. The room, a student's personal environment. In R. Gutman (Ed.), *People and buildings*. New York: Basic Books, 1972.

White, M. Interpersonal distance as affected by room size, status, and sex. *Journal of Social Psychology*, 1975, *95*, 241–249.

Worchel, S., & Teddlie, C. The experience of crowding: A two factor theory. *Journal of Personality and Social Psychology*, 1976, *34*, 30–40.

Zimring, C., Evans, G. W., & Zube, E. Dynamic space: Proxemic research and the design of supportive environments. In A. H. Esser & B. B. Greenbie (Eds.), *Design for communality and privacy*. New York: Plenum, 1978.

13

Density, Personal Control, and Design*

Drury R. Sherrod and Sheldon Cohen

High-density environments are often—but not always—uncontrollable environments. This distinction may be the key in understanding the circumstances under which density adversely affects behavior. In addition, it suggests a number of design-related interventions that can increase the controllability and livability of environments.

High density determines the perceived controllability of environments in two ways. First, the close presence of others can restrict and interfere with the attainment of one's goals. Second, when high density involves the close presence of *strangers*, the environment is not only restricting but also unpredictable—a possible source of irritation or surprise—and thus potentially uncontrollable. Under other circumstances, high density may not be perceived as uncontrollable at all. For example, if goal attainment is not an important issue or if the others present are not strangers, people may experience no loss of control in high-density environments.

The distinction between the effects of density, on the one hand, versus those of control and predictability of environments, on the other, is helpful in clarifying the confusing results of crowding research. Studies of density by itself have found few effects on human behavior.

*This paper was originally presented at the meeting of the Environmental Design Research Association, Vancouver, British Columbia, Canada, 1976.

Drury R. Sherrod • Department of Psychology, Pitzer College, Claremont, California 91711. **Sheldon Cohen** • Department of Psychology, University of Oregon, Eugene, Oregon 97403. Preparation of this paper was supported by a grant from the National Science Foundation (SOC 75-09224).

In contrast, studies of uncontrollable environments have produced a variety of ill effects on human behavior. Thus, it may not be high density *per se* but only uncontrollable high density that is responsible for the negative effects popularly associated with crowding and sometimes observed in research on human density. In a practical sense this could be a very fortunate finding. Although high density is no doubt an unavoidable fact of life for most urban dwellers, controllability is a perceived relationship between self and environment. As a perceptual or cognitive phenomenon, controllability can be altered and fostered by a variety of cognitive, social, and environmental factors. As a result, the effects of "uncontrollable" density can be reduced without reducing density itself. The present chapter is an attempt to distinguish between density and controllability in the research literature. In addition, we hope to show how controllability of environments can be increased by taking into account user needs at several stages in the design process.

Research on Human Density

Research on the effects of human density is beginning to accumulate. Although differences in density manipulations, measures, and mediators make the findings difficult to compare, tentative conclusions are possible. One conclusion based on numerous correlational and experimental studies is that density by itself has fewer and smaller effects on human behavior than anyone expected (see Cohen, Glass, & Phillips, 1979). Correlational studies that control for education and income level find no consistent relationship between external—neighborhood—density and any social pathology, such as rate of crime, disease, mental health, or social disorganization (e.g., Galle, Gove, & McPherson, 1972; Freedman, Heshka, & Levy, 1975; Gillis, 1974; Booth, 1975). Similarly, the majority of well-controlled correlational studies that examine internal density, measured in terms of number of persons per room, also find no relationships or only negligible relationships with social pathologies (e.g., Schmitt, 1966; Freedman *et al.*, 1975; Winsborough, 1965; Mitchell, 1971; Booth, 1975; Gillis, 1974; Levy & Herzog, 1974). Overall, then, these studies, which primarily deal with families residing in high-density dwellings, have found no negative effects of density strong enough to register consistently on measures of crime, physical or mental health, and social disorganization.

Do these findings of well-controlled correlational studies suggest that density is unlikely to affect human behavior adversely in any context? We think not. When the focus of researchers shifts from general populations to specific subgroups low in environmental control, density effects do emerge. Also, laboratory studies suggest that high density does exert negative effects on behavior in certain situations and on certain kinds of tasks.

In contrast to the studies of residential density discussed above, research dealing with population subgroups low in environmental control has revealed pathological effects of internal density. Thus Cohen *et al.* (1979) argue that the evidence for pathological effects of high density is largely limited to such populations as the young (e.g., Booth, 1975), the lower class (e.g., Mitchell, 1971), ship crews (Dean, Pugh, & Gunderson, 1975), and prisoners (e.g., D'Atri, 1975)—all groups with limited environmental control. Similar results have also been reported for college dormitory residents who were tripled in two-person rooms (e.g., Aiello, Epstein, & Karlin, 1975b) and for those who resided on long, undifferentiated (corridor-design) hallways, who presumably lack control over their social interactions (e.g., Baum & Valins, 1977).

Recent laboratory research also suggests that high density can have detrimental effects on human task performance, despite Freedman's (1975) well-publicized argument to the contrary. Even though several studies have found no effects of short-term high density on *simple* task performance (Freedman, Klevansky, & Ehrlich, 1971; Sherrod, 1974; Worchel & Teddlie, 1976; Evans, 1978; Rodin, 1976; see Chapter 5, this volume), two recent studies have shown that density can adversely affect *complex* task performance. Evans (1978) found that subjects in a high-density laboratory setting performed less well on a dual information processing task than low-density subjects, and Paulus and his colleagues (Paulus, Annis, Seta, Schkade, & Matthews, 1976) demonstrated that high density interferes with human maze learning. In addition, several studies have found that high density produces negative aftereffects on a measure of frustration tolerance (Sherrod, 1974; Evans, 1975; Dooley, 1978). Short-term high density has also been found to increase subjects' skin conductance levels (Aiello, Epstein, & Karlin, 1975a). Finally, when high density has involved invasion of subjects' personal space, laboratory subjects have solved fewer anagram puzzles (Worchel & Teddlie, 1976) and have been more affected on aftereffects measures of frustration tolerance (Dooley, 1978) and creativity (Aiello, DeRisi, Epstein, & Karlin, 1977) than high-density subjects whose personal space was not invaded. Thus, while it is clear that short-term high density does not affect simple task performance in the laboratory, high density does seem to affect complex task performance, to reduce postcrowding frustration tolerance and creativity, and to interfere with verbal problem solving when combined with personal space invasion.

Cognitive Factors in Human Density

What do the above situations have in common that may account for the effects of high density? We believe that they are all situations in which controllability is important. The laboratory studies provide a clear demonstration of our point. Strangers in a crowded laboratory may be perceived as unpredictable and thus potentially uncontrollable. According to a recent theoretical paper by Cohen (1978), the unpredictability of others is likely to impede complex but not simple task performance. Cohen argues that unpredictable and uncontrollable environments require close monitoring so that individuals may protect themselves from potential threat or surprise. Such monitoring of the environment demands attentional capacity that would otherwise be available for high information processing needs, such as complex task performance. Thus unpredictability may disrupt complex task performance but not affect performance on less demanding tasks. Similarly, strangers who are invading an individual's personal space also increase the unpredictability of an environment, perhaps more seriously than strangers merely in close proximity but not invading one's personal space. Consequently, the unpredictability of personal space invasion may impede performance on a cognitively demanding task such as verbal problem solving.

Crowded environments also have a second effect, as we noted at the outset of this chapter. Crowding not only increases unpredictability of environments but also restricts freedom and constrains behavior, in effect producing a sense of helplessness. According to Seligman's (1975) theory of learned helplessness, uncontrollability diminishes motivation by setting up an expectancy that an individual's responses are independent of outcomes—in other words, that responses do not matter. As a result, organisms emit fewer responses, and this tendency can generalize to subsequent situations. Consequently, high density may create a sense of learned helplessness that could influence responses on postcrowding measures that are sensitive to motivational defects, for example, a measure of frustration tolerance.

Each of the three performance situations in which high density has produced negative effects—i.e., complex task performance, personal space invasion, and postcrowding frustration tolerance—may therefore be interpreted as resulting from environmental unpredictability or uncontrollability rather than density *per se*. Two studies provide direct support for this interpretation. In the first, Rodin (1976; see Chapter 5) demonstrated that children from high-density environments performed in a laboratory setting as if they were "helpless." Specifically, children from apartments with high person-per-room density were less likely to exercise their own choices in a laboratory free-response situation and were more likely to opt for experimenter-selected rewards instead of self-selected rewards than were subjects from low-density apartments. In addition, children from high-density apartments made more errors in a puzzle-solving task and were more adversely affected by initial exposure to an unsolvable puzzle than were children from low-density apartments. If we assume that children in high-density homes experience more restrictions on their freedom of choice, then these children may have developed little sense of personal control over the environment. Conversely, children from low-density homes may have experienced fewer restrictions of freedom and therefore acquired a greater sense of personal control over the environment. Density, then, may influence persons' general expectations of environmental control.

In the second study that directly supports our position, Sherrod (1974) showed that a perception of control can reduce the aftereffects of short-term laboratory crowding. In this study, subjects were exposed to either high or low density. Some of the high-density subjects were informed that they could leave the crowded room whenever they chose in order to work in another less crowded room. Although no subjects actually left the crowded room, subjects with perceived control over crowding performed better on postcrowding measures of frustration tolerance than did subjects who had no control.

Indirect support for the assertion that perceived control can reduce the effects of crowding can be found in several other studies. Schopler and Walton (1974) found that Internals (i.e., people who generally feel in control of themselves and their environments) felt less crowded in a high-density setting than did Externals (i.e., people who feel controlled by the environment). Similarly, Karlin, Epstein, and Aiello (1975) found some indications that Internals were less influenced than Externals by group processes in a crowded environment. Also, Baum and Valins (1977) found that members of cohesive groups felt less crowded in high-density dormitories than people who are not members of cohesive groups. If we assume that cohesive groups allow for greater predictability and consequent controllability of one's environment, then this study fits in well with our perspective. Finally, Worchel and Teddlie (1976) found that when crowded people are distracted from attending to the close proximity of other persons by the presence of attractive wall posters, they experience the environment as less crowded and perform better on verbal puzzle-solving tasks. In this experiment, when subjects' attention was diverted from the presence of other people in the crowded environment, it is possible that subjects then felt less concerned about the issue of control than did subjects whose attention was not distracted. Each of these studies then provides some indirect support for our perspective regarding human density and control.

Effects of Personal Control

Controllability also has independent effects on human behavior above and beyond its effects in mediating reactions to density. The belief in one's ability to control the environment has been clearly shown to have a variety of important psychological effects. In their work on urban stress, Glass and Singer (1972) found that individuals who had some perception of control over loud noise, electric shock, or frustrating bureaucracies performed better on measures of frustration tolerance and attention to detail than did subjects who had no perception of control, even though none of the subjects with perceived control ever actually exercised their options and escaped the environmental stressors.

From another perspective, Seligman and his colleagues (cf. Seligman, 1975) have demonstrated that both animals and humans tend to give up in a free-response situation if they have experienced prior uncontrollable aversive stimulation. Not only do such individuals tend to give up when their continued responses could actually bring about relief from stress, they even fail to learn when environmental contingencies have changed and their responses could finally be successful. In contrast, when individuals are exposed to prior escapable aversive stimulation, they easily learn that subsequent situations are also escapable and they quickly emit the necessary responses.

In still another approach to the control issue, de Charms (1968) has argued that uncontrollable situations cause individuals to see themselves as "Pawns" of the environment, whereas persons who feel in control of the environment come to see themselves as "Origins." A self-perception as an Origin or a Pawn affects one's general sense of competence and motivation for behavior. Pawns respond with passivity across a variety of situations as if they were helpless, whereas Origins emit a high level of voluntary responses, in the belief that they can successfully manipulate the environment. These theories suggest that a sense of controllability not only may reduce the effects of uncontrollable high density, but also may serve to alter an individual's expectations about the value of voluntary responses and to influence one's self-perceptions as a competent human being.

Personal Control and the Design Process

If the perceived relationship between self and environment can produce such profound consequences in the laboratory, a relevant question then becomes: To what extent is it possible to increase people's perceptions of controllability in real-world settings? As we noted at the outset of this chapter, although crowding is often an unavoidable physical fact, perceived control is an alterable cognitive phenomenon. It is important to emphasize that the belief in one's ability to control the environment need not imply the ability to actually implement control. Perceived control *may* result from actual control, but it can also result from prior control experiences, from information suggesting that control is potentially available, from self-inferences, or from any social or physical intervention that makes the environment appear more manageable or predictable. Perceived control may even be illusory, but its effects on human behavior can be significant, as numerous experiments have shown.

We believe that an individual's sense of control over the environment can be enhanced in a variety of ways. Environments not only can be physically designed so as to make them more controllable, they also can be made to *appear* more controllable as a result of social and cognitive interventions.

A major issue in designing more controllable environments is a problem that can be labeled "environment–function fit." Simply put, when the environment fits well with the functions to be performed there, the environment is manageable; if there is not a good fit, the environment is less manageable, less controllable. To determine environment-function fit, designers must take into account the demands of the tasks and behaviors to be performed in the environment. For example, what will be the effect of environmental distractions on task performance? How important is social support and communication, and to what extent does the environment allow for this? Although it is impossible to outline all relevant factors for any particular environment, the essential point is that an advance analysis be made of the task functions and social functions to be performed in the environment. Manageable environments are predictable and controllable whereas "nonfitting" environments deprive one of perceptions of control.

The best way of assessing potential environment–function fit is to go directly to the users—not the administrators and managers, but the users. Not only should uses and patterns of behaviors be observed and surveyed, but the users should also be directly involved in planning and specifying the design needs. User participation in the design process allows a sense of personal control over the environment. Even if users choose not to participate, they can still benefit from the opportunity. For example, in some versions of Glass and Singer's experiments on perceived control, it is noteworthy that subjects who merely had access to a person with control over noise were as unaffected by noise as were those subjects who had control themselves. Finally, in order to completely assess environment–function fit, users should be consulted after occupancy to see how well the environment actually meets their needs. Besides providing information that could result in improving the user's relationship to his environment, this extended participation in the design process serves to reinforce their perceptions of control.

When space is limited and it is impossible to build a separate environment for each function to be performed, flexibility becomes important. Altman (1975) has recently emphasized the importance of "responsive environments," or environments that can be altered to suit different needs—in particular, for either privacy or togetherness. However, a cautionary note should be sounded here in light of Seligman's helplessness research. If users have never before inhabited flexible environments, they may feel relatively helpless regarding their ability to alter the environment. As a result they may fail to realize or to exercise their environmental options. Therefore, it would be important to take into account the users' existing behavioral repertoire and expectations and perhaps to provide users with special training where established patterns of behavior do not already exist. In other words, objective flexibility may not increase users' perceptions of control unless users know how to use their new freedoms and perceive these new alternatives as useful.

Other design factors relate more specifically to high density. As we pointed out earlier, density can increase unpredictability in environments and can create feelings of helplessness by restricting behavioral options. Even though unpredictability is probably not a significant factor within the home, where others are generally well known, density may nevertheless produce feelings of helplessness, as demonstrated by Rodin's experiment. Perhaps a design feature that could increase feelings of control in residential environments would be to build "escape rooms"—private nooks or enclosed modules—when economically feasible. Such a space within an apartment could allow a respite or retreat from social interaction. This is one of the functions that the bathroom plays in many homes. Why not design a more appropriate space specifically for this purpose? Even when the space is not in use, its mere availability could ameliorate some effects of density. As demonstrated by Sherrod's experiment, escapable crowding can increase postcrowding frustration tolerance even though escape may never be exercised. Similar private spaces might be included in apartment houses or dormitories in the form of small, private rooms that residents could reserve for reading, writing, or just a time to be alone. Again, it is less important that the space be used than that it is simply available.

Outside the immediate family, unpredictability becomes a factor that can influence the effects of crowding. For example, the greater the number of tenants in a building, or the more apartments on a corridor, or the more people who must use a recreational area, the less likely are residents to know each other well and the more unpredictable are the others to whom one is exposed. However, friendship formation and predictability can be increased by building smaller apartment houses, breaking up long corridors, and decreasing the number of people served by a recreational area. Such practices would foster community, decrease anonymity, and make environments more predictable and controllable. In addition, by encouraging the development of cohesive groups, such environments may actually cause people to feel less crowded, as demonstrated by the research of Baum and Valins.

Other features can be incorporated into the larger urban environment to increase perceptions of control. As Lynch (1960) has argued, urban environments are more "codable" when they include such salient features as well-defined neighborhoods, landmarks, points of interest, and clear pathways. These features allow individuals to form clearer cognitive maps of an area, facilitate information processing, increase predictability, and enhance perceptions of control. In addition, as Jacobs (1961) has argued, when buildings and communities are designed to encourage the use of streets and sidewalks, a sense of community develops and the streets become safer. From our own persepctive, when the streets are friendlier and safer, they are less threatening and more predictable.

In summary, we have suggested a number of ways that predictability and controllability of environments can be increased. The research suggests that greater perceptions of controllability can influence a wide range of behaviors. Moreover, controllability may be the single most important factor that mediates the effects of crowding. As we have argued throughout this chapter, the effects of crowding depend upon the behaviors being performed in an environment as well as upon the individual's appraisal of the environment as controllable or uncontrollable. In a real sense, the most important determinant of human behavior may not be the physical environments we inhabit but the cognitive environments we perceive.

ACKOWLEDGMENTS

The authors are indebted to Dan Stokols for his comments on an earlier draft.

References

Aiello, J. R., DeRisi, D. T., Epstein, Y. M., & Karlin, R. A. Crowding and the role of interpersonal distance preference. *Sociometry*, 1977, 40, 271–282.

Aiello, J. R., Epstein, Y. M., & Karlin, R. A. Effects of crowding on electrodermal activity. *Sociological Symposium*, 1975. (a)

Aiello, J. R., Epstein, Y. M., & Karlin, R. A. *Field experimental research on human crowding*. Paper presented at the meeting of the Eastern Psychological Association, New York City, 1975. (b)

Altman, I. *The environment and social behavior*. Monterey, California: Brooks/Cole, 1975.

Baum, A., & Valins, S. *Architecture and social behavior: Psychological studies of social density*. Hillsdale, New Jersey: Erlbaum, 1977.

Booth, A. Final report: Urban crowding project. Ministry of State for Urban Affairs, Government of Canada, August 1975.

Cohen, S. Environmental load and the allocation of attention. In A. Baum & S. Valins (Eds.), *Advances in environmental research*. Hillsdale, New Jersey: Erlbaum, 1978.

Cohen, S., Glass, D. C., & Phillips, S. Environmental factors in health. In H. E. Freeman, S. Levine, & L. G. Reeder (Eds.), *Handbook of medical sociology*. Englewood Cliffs, New Jersey: Prentice-Hall, 1979.

D'Atri, D. A. Psychophysiological responses to crowding. *Environment and Behavior*, 1975, *7*, 237–252.

Dean, L. M., Pugh, W. M., & Gunderson, E. Spatial and perceptual components of crowding: Effects on health and satisfaction. *Environment and Behavior*, 1975, *7*, 225–236.

deCharms, R. *Personal causation*. New York: Academic Press, 1968.

Dooley, B. B. Effects of social density on men with "close" or "far" personal space. *Journal of Population*, 1978, *1*, 251–265.

Evans, G. W. Behavioral and physiological consequences of crowding in humans. *Journal of Applied Social Psychology*, 1978, in press.

Freedman, J. L. *Crowding and behavior*. San Francisco: Freeman, 1975.

Freedman, J. L., Klevansky, S., & Ehrlich, P. R. The effect of crowding on human task performance. *Journal of Applied Social Psychology*, 1971, *1*, 7–25.

Freedman, J. L., Heshka, S., & Levy, A. Population density and pathology: Is there a relationship? *Journal of Experimental Social Psychology*, 1975, *11*, 539–552.

Galle, O. R., Gove, W. R., & McPherson, J. M. Population density and pathology: What are the relationships for man? *Science*, 1972, *176*, 23–30.

Gillis, A. Population density and social pathology: The case of building type, social allowance, and juvenile delinquency. *Social Forces*, 1974, *53*, 306.

Glass, D. C., & Singer, J. E. *Urban stress: Experiments on noise and social stressors*. New York: Academic Press, 1972.

Jacobs, J. *The death and life of great American cities*. New York: Random House, 1961.

Karlin, R. A., Epstein, Y. M., & Aiello, J. R. *The effects of internal versus external locus of control on reactions to crowding*. Unpublished manuscript, Rutgers University, 1975.

Levy, L., & Herzog, A. N. Effects of population density and crowding on health and social adaptation in the Netherlands. *Journal of Health and Social Behavior*, 1974, *15*, 228–240.

Lynch, K. *The image of the city*. Cambridge, Massachusetts: M.I.T. Press, 1960.

Mitchell, R. E. Some social implications of high-density housing. *American Sociological Review*, 1971, *36*, 18–29.

Paulus, P. B., Annis, A. B., Seta, J. J., Schkade, J. K., & Matthews, R. W. Density does affect task performance. *Journal of Personality and Social Psychology*, 1976, *34*, 248–253.

Rodin, J. Density, perceived choice and response to controllable and uncontrollable outcomes. *Journal of Experimental Social Psychology*, 1976, *12*, 564–578.

Schmitt, R. C. Density, health, and social disorganization. *Journal of the American Institute of Planners*, 1966, *32*, 38–40.

Schopler, J., & Walton, M. *The effects of structure, expected enjoyment, and participants' internality-externality upon feelings of being crowded*. Unpublished manuscript, University of North Carolina at Chapel Hill, 1974.

Seligman, M. E. P. *Helplessness*. San Francisco: Freeman, 1975.

Sherrod, D. R. Crowding, perceived control, and behavioral aftereffects. *Journal of Applied Social Psychology*, 1974, *4*, 171–186.

Winsborough, H. H. The social consequences of high population density. *Law and Contemporary Problems*, 1965, *30*, 120–126.

Worchel, S., & Teddlie, C. The experience of crowding: A two-factor theory. *Journal of Personality and Social Psychology*, 1976, *34*, 30–40.

14

Designing for High-Density Living

Allen Schiffenbauer

Many people in our culture choose to live in high-density areas and sometimes pay a premium to do so. No one is forced to live in a high-rise apartment on New York's Park Avenue; Chicago's near North Side is exciting and attractive because of, not in spite of, its high density. In some societies it seems as if the whole culture opts for high density. Rapoport (1969) suggests that this is true for Greeks; Draper (1973) reports that the !Kung Bushmen live under crowded conditions that are not dictated by either economic or geographic necessity.

More formally, the experimental psychological literature concerning human responses to crowding produced early evidence (Freedman, Klevansky, & Ehrlich, 1971) that high density does not necessarily have negative effects. Since these early studies, a considerable body of literature has been published.[1] Taken as a whole, this body of literature demonstrates that under certain circumstances high density can be interpreted as the cause of a large catalog of human ills (e.g., poor health, poor grades, withdrawal from social contacts, decrements in task performance, and disruption of social relations), while under different conditions, high density will in and of itself have no negative effects (Lawrence, 1974; Freedman, 1975; Sundstrom, 1978).

---

[1]An in-depth review of the entire body of crowding literature is clearly beyond the scope of this chapter. For an exhaustive review, the interested reader is referred to Sundstrom, 1978.

Allen Schiffenbauer • Research Division, Needham, Harper, & Steers, Inc., Chicago, Illinois 60611. This research was conducted while the author was on the faculty of Virginia Polytechnic Institute and State University.

Once this point is made the next question is, What conditions will make high-density living tolerable? The question should be phrased in this way, because it is clear that certain environments (e.g., dormitories, prisons, center-city apartments) must be designed to accommodate high densities. Since high-density housing *will* be constructed, the problem becomes one of finding designs that will ameliorate negative effects. We have at least two sources of suggestions about what these designs might be like—the architectural literature and the psychological literature.

Architecture and Psychology

Some architects and planners have suggested that high-density designs can be quite pleasant. For Jane Jacobs (1961) the city is made vibrant by the diversity high density makes possible. Le Corbusier's (1929/1971) "radiant city" was designed to incorporate very high densities of people in large high-rises amid beautiful open parks. Le Corbusier was convinced that this sort of arrangement would allow people to lead happy lives both because of and in spite of high densities. Soleri (1969) offers another solution to the problem. He suggests huge megastructures miles in diameter housing millions of people. His drawings of what these megastructures might look like are different from Le Corbusier's parks and remind one of nothing so much as an insect colony. These and other architectural solutions offer compelling images, but as practical solutions they have some problems. Their major weaknesses are in the ways architects think about and deal with the people who will eventually inhabit their structures. In the worst cases, designers show a callous disregard for the fact that people must live in the environments they create. They become more interested in construction techniques or forms manipulation than in the people who will live in their buildings. In many other cases, architects try to be sensitive to the needs of the eventual inhabitants. With all good intentions, they rely on their own intuitions and the untested intuitions of other architects' views of human nature. These intuitions can be painfully misleading for a number of reasons. The architect has a personal and cultural history that is often drastically different from those of the eventual inhabitants. White, university-trained, middle-class designers may not know the world of the person who lives in the public housing projects they design. If the architect experiences a different world from the inhabitant, perhaps the architect's intuitions should not serve as a basis for designing the inhabitant's residence. In addition, when a particular intuition is translated into a design and constructed, there is usually no evaluation of the social psychological success of the building. There is no systematic testing of particular ideas about the behavioral impact of design, and, furthermore, the psychologist does not usually include any specific design implications of his work. Designers may have some notion that what they are reading is relevant to their work, but they receive little help from the psychological source in translating the behavioral findings into design. Even when an article does draw design implications, they may be so naive concerning the economic and technological constraints designers labor under that they may reject the behavioral resources.

Many roadblocks in the way to successful collaboration arise because the architect and the psychologist come from two cultures that differ in many ways. For example, scientists are engaged in the pursuit of truth at a sometimes leisurely pace. They are oriented toward the testing of abstract general concepts, while working architects are involved with a specific, concrete set of problems unique to their current project—two different orientations that do not mesh easily. Another "cultural difference" is that scientists are conservative. They are often unwilling to make recommendations about new, complex, "real-world" situations that "go beyond" their data without preparing a few experiments. This conservatism frustrates the architect, who cannot afford the luxury of the scientist's pace.

This difference poses a problem for psychologists. If they exercise their judgment and go beyond their data to suggest some concrete designs that they have *not* tested, they may satisfy the architect, but they run the risk of offending their colleagues and perhaps some of their own values. If, however, they delay making any design suggestions until they are sure of themselves by rigorous scientific standards, they will never be able to be useful within a designer's normal time framework.

Not surprisingly, the psychological and architectural professions present practitioners with different reinforcement contingencies that dictate different sorts of work if professional advancement is to be forthcoming. Architects win awards and get their designs published on the basis of the way their buildings "look." A brief review of architectural journals reveals an abundance of pictures and a paucity of evaluation. As James Marston Fitch has noted, much architectural criticism has been written by people who have never seen the buildings they write about, only pictures of them. For their part, academic psychologists are bound by many pressures to eschew applied work, and to present their results in a traditional format. As the architect's journals communicate in the visual language of that profession, so the psychologists use the words and numbers their colleagues understand. In much the same way that the psychologist is struck by the visual orientation of the architect, the architect notices the lack of photographs, drawings, or floor plans in the psychologist's work. This observation, and the ones preceding, are not intended to be an exhaustive list of all the problems confronting the integration of social science data via design. It is intended to point out the fact that interdisciplinary communication, especially when confounded by an applied–basic split, is not easy and will not take place just because it should.

The research reported here was designed with the dual purposes of producing data that would further the scientist's conceptual understanding of the effects of high-density living and information that architects could use in the design process. To make the data relevant to designers' concerns, several steps were taken. First, the experiments were performed *in vivo* rather than in laboratory settings. Often the designer is concerned about how well laboratory research can be generalized to real-world settings. Doing the research in the real world to begin with gives the results an ecological validity that is necessary if the architect is going to base decisions on them. Another feature of this research is that the conceptual variables important to the scientist are operationalized and discussed in terms of the architectural features over which the designer has control. Finally, an attempt is made to suggest some design alternatives incorporating the findings of the research.

Study I

Whether or not high density is correlated with crowding[2] or any other negative effects (e.g., infant mortality, fertility, welfare rate, juvenile delinquency rate) seems to be at least partially dependent upon how density is defined. If density is defined in terms of dwelling units per acre, or dwelling units per structure, then density seems to have little or no negative effect. If, however, density is defined in terms of persons per room, then the effects of density can be marked (Galle, Gove, & McPherson, 1972). One way of explaining the importance of this difference definition is Stokols's (1976) differentiation between primary and secondary environments.[3]

To return briefly to an example mentioned above, Park Avenue apartment-dwellers might very well experience high density in many of their relatively unimportant secondary environments. The subways, elevators, and streets they use might be overpopulated, but their homes

---

[2]From this point forward the author will follow the distinction first proposed by Stokols (1972) that density be used to refer to physical conditions, whereas the term *crowding* be used to refer to a psychological state.

[3]A primary environment is one in which individuals spend much time and one in which they perform important activities. A secondary environment is one in which individuals spend far less time and perform less important activities.

need not be. No matter what the quality of the secondary environments they experience, they can always retreat into the comparative solitude of their own homes—their primary environments. By retreating into their private, primary environments, individuals can accomplish several things; among them (1) they can limit the number of people with whom they must interact; (2) they can limit the number of social interruptions they are likely to experience; (3) they can experience a sense of control over their environment. Note, however, that our mythical apartment-dwellers can accomplish these things only if, as well as retreating, they can also *prevent others from entering their primary environments.* That is, a private place can be a private place only if access to it by others is controlled.

In terms of the psychological literature, there is very good reason to believe that the functions served by retreat are intimately tied to the experience of crowding and other density effects. Sundstrom (1978), after an extensive review of the literature, concludes that interference and the number of others with which one must interact are two of the three (along with immediacy) critical factors in density effects. As for the third function of retreat, the role of control in ameliorating the impact of negative events, in general, has been clearly demonstrated (Averill, 1973; Zimbardo, 1969). Laboratory evidence bears more directly on this point. Feelings of control have been found to lessen the impact of high density (Sherrod, 1974). The study reported below shows that feelings of loss of control over the environment play a role in the impact of density even in field settings.

Schiffenbauer (1976c) studied the reactions of 53 college freshman to their dormitories during their first 6 weeks on campus.[4] The author had hoped to study the effects of room size while holding interaction rate constant by using only those students who had been randomly assigned to double rooms as subjects. This control turned out to be difficult. Although the number of persons assigned to each room was always the same, the number of visitors to each room varied. Furthermore, this particular variable, amount of visitation, was strongly related to crowd-ing and several other variables. Students who had many hours of daily visitors tended to feel more crowded, reported difficulty in studying, did not like their roommates, were not satisfied with their rooms, and, not surprisingly, were annoyed by visitors (see Table 1). The degree to which students were annoyed by visitors was also related to the amount of *control* they felt over the room; subjects who were annoyed most and suffered the most disruption felt the least control over their own rooms. It would seem that, in these dormitories, students who retreat to their

---

[4]For a more detailed description of the procedure used see the Appendix.

Table 1.   Correlations with Number of Hours per
Day a Student Is Visited ($df = 51$)

Variable	Correlation
Perceived crowding	.32[a]
Difficulty in studying	.72[b]
Liking for roommate	−.79[b]
Satisfaction with room	−.77[b]
Annoyance with visitors	.64[b]

[a] $p < .02$.
[b] $p < .001$.

rooms do not necessarily achieve what our Park Avenuer does. At least some students find themselves inundated with unwanted visitors who interrupt them, make them interact when they do not wish to, and make them experience a loss of control over their environment.

The correlations presented in Table 1 yield a very consistent pattern. Students who are visited too often suffer a variety of negative consequences; they feel crowded, have difficulty studying, dislike their roommates, and are dissatisfied with their rooms. While one must always exercise caution when inferring causal relationships from the sort of correlational data reported in this study, it seems likely the amount of visiting is the causal variable while the others are effects. It also seems reasonable to suggest that crowdedness and loss of control are crucial moderating variables.

If excessive visiting leads to feelings of crowdedness and other unwanted effects, then any way of decreasing the amount of unwanted visitation should reduce or eliminate the negative effects. The question is, how might unwanted visitors be discouraged? More specifically, is there any way in which the physical environment might be designed so as to discourage unwanted visitors?

Several architects have attempted to address this question. Chermayeff and Alexander (1963) state the problem outlined above very evocatively:

> What was once commonplace—the possibility of escape from the crowd for privacy and rest—has all but vanished. The crowds, once restricted to the streets and borders of the public domain, now follow unbidden into the solitary, private domain. (p. 74)

To solve this problem, "The individual requires barriers against the sounds and sight of innumerable visitors . . . (Chermayeff & Alexander, 1963). They go on to recommend that people be provided with some sort of buffer between their private spaces and the public world outside. In other terms, people would be protected from unwanted intrusions into their dormitory rooms if there was some sort of semiprivate space between the room and the public corridor (see Newman, 1972, for a discussion of the differentiation of public, private, semipublic, and semiprivate spaces).

At present, students walking down the public corridors feel free to intrude into a closed room. Clearly, the door in itself does not offer enough of a barrier to discourage unwanted social contact. Students are aware of this problem. When asked how the physical environment might be changed to make it more livable, one male student said, "Wire the door knob for electric shock." If the space in front of each student's door were differentiated from the rest of the corridor in some clear way, then a semiprivate buffer would be created. In suite designs, this end is often accomplished by the presence of a shared sitting room placed between the traffic-bearing corridors and the student's sleeping quarters (Baum & Valins, 1973). In the more common double-loaded corridor design, no such differentiation is provided. A first step in carrying out differentiation in a double-loaded corridor might be to change the texture or color of the flooring immediately in front of each student's room. Props might also help. A door mat might provide level and texture changes simultaneously. Although a narrow double-loaded corridor offers little room for experimentation, there are opportunities to provide students with spaces that might act in ways analogous to the suite sitting room, the front porch of a rural home, or the stoop of an urban town house. This solution is analogous in the sense that all of these spaces act as semiprivate buffers that help to regulate intrusions.

The differentiation would "warn" students approaching a door that they are about to impinge upon others' private space and make them more thoughtful about doing so. (In the language of behavior modification, the differentiation acts as a prompt.) This prompt should decrease the probability that visitors will enter a room uninvited. This sort of solution can be seen as decreasing visitation by increasing resident control. At present, the door does not provide adequate control. Opening the door clearly invites interaction, but closing it does not clearly discourage it. Providing a semiprivate buffer should increase the residents' ability to control their social contacts by making visitors more aware of the fact that they are about to intrude where they may not be wanted.

To this point, the only type of control that has been mentioned is control over social interaction. Control over the physical environment seems to relate to behavior in much the same way as control over the social environment. Students who felt they could arrange their furniture in many different ways tended to feel less crowded ($r = .35$, $df = .51$, $p < .01$), liked their roommates better ($r = .36$, $df = .51$, $p < .01$), and tended to be more satisfied with their rooms. The implications of these findings are clear: if one wants to decrease the effects of density, the environment should be flexible enough to allow the residents to exercise some control over their environments. Hard architecture (Sommer, 1974) should be abandoned in favor of softer, more responsive designs.

The exercise of control over the environment may influence reactions to it in two ways. First, if the users are allowed control, they may design their spaces and experiences so that they better suit their needs. This is the level at which most designers consider user control. There is, however, another aspect of control. Several authors (Brehm, 1966; Zimbardo, 1969; Averill, 1973) have emphasized the fact that feelings of control, in and of themselves, are important in modifying the impact of potentially aversive stimulation. Translated into design terms, this means that if users exercise control over their environments, either in design or management, they will experience the environments more positively, *regardless of the quality of the user's solution*.

This finding suggests that management techniques can serve the same ends as design manipulations. For example, allowing students to decorate their own rooms might give them feelings of control, allowing personalization should facilitate beneficial identification with their adopted primary environments.

Study II

This section will present an examination of the relationship between design and crowding in one dormitory and will discuss some ways in which this relationship can be modified by the design of the dormitory room.

Often the physical size of a building and the general size of the interior rooms are determined by financial and site considerations. In addition, the number of individuals a building must house is often a fixed part of the program with which the architect must work. If square footage and number of resident users is fixed, then the gross density of the building is also fixed. In dormitories and most urban housing, this density is usually high. Within these constraints, it is the architect's task to find a sensitive solution to the problems created by high density.

The data reported below were primarily collected in Slusher Tower; a 12-story, all-female dormitory on the Virginia Polytechnic Institute and State University campus. Each floor above ground level is composed of two suites of six rooms each, with two women assigned to each room. The women from both suites share common bathroom and lounge facilities. The building was designed so that all rooms are the same size

(9 × 16 feet). Since each room is the same size, and each room has the same number of residents, the objective physical density of each room is the same—72 square feet per person. Each room is supplied with two beds and a built-in wall unit that contains two closets, mirrors, dressers, and desks. The residents can add furniture if they wish, and they often do add an easy chair or carpet. The rooms are so small, however, that the amount of student-contributed furniture is always small.[5]

Even though all the rooms are of identical physical size and tend to have the same furniture, there is large variability in how large and how crowded students feel their rooms to be. This variability is systematically related to certain physical features of the individual dormitory rooms.

In one study (Schiffenbauer, Brown, Perry, Shulack, & Zanzola, 1977) the floor that the student lived on was significantly related to how large the students judged their rooms to be; students who lived on the higher floors tended to feel that their rooms were larger. Floor was not related to how crowded students thought their rooms were. A second study (Schiffenbauer, 1976c) replicated the room size finding and found that students who lived on higher floors felt less crowded. Floor was also related to roommate interactions; those subjects who lived on higher floors tended to get along with their roommates better. Since the subjects in both studies were randomly assigned to their rooms by the housing office, these findings seem to suggest that something about living high up allows students to feel they have more space and are able to get along better with their roommates.

The reason for this relationship has not been firmly established. Schiffenbauer et al. (1977) suggest that the higher up one is, the more visually expanded the environment. People in the room can see further when they look out the window, and this cognitive/perceptual expansion makes them feel and act as if the room were larger than it actually is. They see the floor effect as analogous to the way in which a mirrored wall can make a small room appear and "feel" larger. Others (see Sundstrom, 1978) suggest that the windows in the higher floor rooms provide more visual escape than do the windows in the lower floor rooms. Given the fact that in this particular case the window placement is identical in all the rooms, the expansion hypothesis seems more reasonable than the escape notion.

The environment might be visually expanded in other ways to achieve similar effects. Increasing the number of windows, placing windows so as to get the most expansive view, the use of mirrors, and the

---

[5]Some of these data are reported in more detail in Schiffenbauer et al. (1977). The interested reader will also find a more complete description of the procedure used in the same manuscript.

use of graphics, all represent relatively cheap ways of expanding the environment. Perhaps the students living on the upper floors feel that they have more control over their environment because of lower traffic flow in these areas, or possibly lower traffic volume might lead to a greater sense of defensibility since fewer students pass through (Newman, 1972).

The brightness of the room also affected its perceived crowdedness. In two studies (Schiffenbauer, 1976b; Schiffenbauer et al., 1977), it was found that people who judged their rooms to be light also judged their rooms to be uncrowded. In another study (Schiffenbauer, 1976c), it was found that those rooms judged light in previous studies also contained roommates who got along better with each other. Once again, a pattern of results emerges in which some variable, light in this case, interacts with physical density to affect feelings of crowding and, perhaps more importantly, the consequences of crowding.

The fact that light rooms seem less crowded and foster improved roommate interaction has clear implications for design. Light levels should be kept reasonably high; room colors should be light. Two cautions should be kept in mind when attempting to apply these findings: (1) in all cases, the light levels were levels of sunlight. Since the quality of sunlight can differ markedly from that of artificial light, some caution should be exercised in cases where light levels are manipulated with artificial light. (2) These studies measured the effects of light over a restricted range of values. It is almost certainly true that the relationship between light level and user satisfaction is curvilinear; very high levels of light, especially if the light produces glare, would be aversive.

Conclusion

The data presented above demonstrate that it should be possible to reduce the negative effects of high-density living through the sensitive manipulation of architectural features. These data suggest two quite different ways designers can go about designing livable high-density housing. First, they can try to design their spaces in such a way that they are perceived as roomy and uncrowded (e.g., by making bright and visually expanded rooms). Second, they can provide residents with the ability to manage the environment so that it is responsive to their needs (e.g., give residents the ability to control their social interaction rate).

The control issue is of particular interest for several reasons: (1) A growing body of literature indicates that perceived control may be the central mediating factor in crowding. (2) The control question also points out the fact that a designer should consider the way in which the environment is managed as well as the way in which it is constructed. Often architects think of the buildings they create as static entities. They ignore the fact that the building space is also a life space for the people who live there. Although the building itself may be static, the activities that it encloses certainly are not. If an environment is to be successful, the architect must provide some way for the inhabitants to manage their space to conform to their needs. People will act on the environment to bring it into line with their life-styles (e.g., Boudon, 1972) whether the architect likes it or not. Therefore, the issue for the architect becomes how to make the environment as responsive as possible. In the case of high-density living arrangements, the architect must provide individuals with the ability to manage their social interaction rate. It might seem as if this is an easy issue—just provide a solid door—but, as we have seen in the studies reported above, this simple solution does not always work. Other solutions should be tried.

The analysis presented here is not meant to suggest that control, visual expansion, and brightness will produce a panacea for the ills of high-density housing, nor should anything in this chapter be read as implying that a psychological approach should supplant an architectural one. Even though the data alone cannot generate a design, they can lead to a more informed design by keeping architects constantly aware of the fact that people will be inhabiting their buildings—people with needs that must be met—and the psychological literature can give them guidance.

Appendix

Method

Subjects. Subjects were 53 Virginia Polytechnic Institute and State University freshmen recruited out of an Introductory Psychology class. Subjects participated in the study for extra class credit. All dormitories on campus were represented in the sample except for the special R.O.T.C. and athletic dormitories.

Procedure. All subjects were administered a series of questionnaires once each week for the first six weeks of their residence on campus. Subjects were always tested in one group at the same time and place each week.

The questionnaires administered included several standard personality tests as well as questions designed to tap students' feelings about their rooms and roommates. The hours of visits per day measure was assessed by subject self-reports.

References

Averill, J. R. Personal control over aversive stimuli and its relationship to stress. *Psychological Bulletin*, 1973, *80*, 286–303.

Baum, A., & Valins, S. Residential group size, social interactions, and crowding. *Environment and Behavior*, 1973, *5*, 421–439.

Boudon, P. [Lived-in architecture: Le Corbusier's Pessac revisited] (G. Onn, Trans.). Cambridge, Massachusetts: M.I.T. Press, 1972.

Brehm, J. W. *A theory of psychological reactance.* New York: Academic Press, 1966.

Chermayeff, S., & Alexander, C. *Community and privacy: Toward a new architecture of humanism.* New York: Doubleday, 1963.

Draper, P. Crowding among hunter-gatherers: The !Kung bushmen. *Science*, 1973, *182*, 301–303.

Freedman, J. L. *Crowding and behavior.* New York: Viking Press, 1975.

Freedman, J. L., Klevansky, S., & Ehrlich, P. The effect of crowding on human task performance. *Journal of Applied Social Psychology*, 1971, *1*, 7–25.

Galle, O. R., Gove, W. R., & McPherson, J. M. Population density and pathology: What are the relations for man? *Science*, 1972, *176*, 23–30.

Jacobs, J. *The death and life of great American cities.* New York: Random House, 1961.

Lawrence, J. E. S. Science and sentiment: Overview of research on crowding and human behavior. *Psychological Bulletin*, 1974, *81*, 712–720.

Le Corbusier. *The city of tomorrow.* Cambridge, Massachusetts: M.I.T. Press, 1971. (Originally published, 1929.)

Newman, O. *Defensible space: Crime prevention through urban design.* New York: Macmillan, 1972.

Rapoport, A. *House form and culture.* Englewood Cliffs, New Jersey: Prentice-Hall, 1969.

Schiffenbauer, A. I. *Issues in interdisciplinary collaboration between psychologists and architects.* Paper presented at American Psychological Association, Washington, D.C., 1976. (a)

Schiffenbauer, A. I. *The relationships among social density, spatial density and crowding in a dormitory setting.* Unpublished Manuscript, Virginia Polytechnic Institute and State University, 1976. (b)

Schiffenbauer, A. I. *The social ecology of Slusher Tower.* Unpublished manuscript, Virginia Polytechnic Institute and State University, 1976. (c)

Schiffenbauer, A. I., Brown, J. E., Perry, P. L., Shulack, L. K., & Zanzola, A. M. The relationship between density and crowding: Some architectural modifiers. *Environment and Behavior*, 1977, *9*, 3–14.

Sherrod, D. R. Crowding, perceived control, and behavior after effects. *Journal of Applied Social Psychology*, 1974, *4*, 171–186.

Soleri, P. *Arcology: The city in the image of man.* Cambridge, Massachusetts: M.I.T. Press, 1969.

Sommer, R. *Tight spaces: Hard architecture and how to humanize it.* Englewood Cliffs, New Jersey: Prentice-Hall, 1974.

Stokols, D. On the distinction between density and crowding: Some implications for future research. *Psychological Review*, 1972, *79*, 275–277.

Stokols, D. The experience of crowding in primary and secondary environments. *Environment and Behavior*, 1976, *8*, 49–86.

Sundstrom, E. Crowding as a sequential process: Review of research on the effects of population density on humans. In A. Baum & Y. Epstein (Eds.), *Human response to crowding.* Hillsdale, New Jersey: Earlbaum, 1978.

Zimbardo, P. *The cognitive control of motivation: The consequences of choice and dissonance.* Glenview, Illinois: Scott, Foresman, 1969.

Author Index

Subject Index